Study of Nuclear and Alternative Energy Systems

U.S. ENERGY SUPPLY PROSPECTS TO 2010

The Report of the
Supply and Delivery Panel
  to the
Committee on Nuclear and Alternative Energy Systems
National Research Council

NATIONAL ACADEMY OF SCIENCES
WASHINGTON, D.C.   1979

International Standard Book Number 0-309-02936-8

Library of Congress Catalog Card Number 79-90990

Available from:

Office of Publications
National Academy of Sciences
2101 Constitution Avenue, N.W.
Washington, D.C. 20418

Printed in the United States of America

PREFACE

In June 1975, the National Research Council (NRC) undertook a comprehensive study of the nation's prospective energy economy during the period 1985-2010, with special attention to the role of nuclear power among the alternative energy systems. The goal of the study was to assist the American people and government in the formulation of energy policy.

The Governing Board of the National Research Council appointed an NRC-wide Committee on Nuclear and Alternative Energy Systems (CONAES) to conduct the study, the committee consisting of 14 members from diverse backgrounds and disciplines.

The Committee developed a three-tiered functional structure for the project, the first tier being CONAES itself, whose report will embody the ultimate findings, conclusions and judgments of the study. To provide scientific and engineering data and analyses for the committee, a second tier of four panels was formed to examine (1) energy demand and conservation, (2) energy supply and delivery systems, (3) risks and impacts of energy supply and use, and (4) syntheses of various models of future energy economies and decision-making. Each panel, in turn, established a number of resource groups -- 22 in all -- as the third tier, to address in detail an array of more particular matters.

The committee asked the Supply and Delivery Panel to determine the extent to which domestic supplies of energy could be produced, should that production become desirable. Specifically, the panel investigated

o The availability of primary energy resources in the United States, and the prospects for their recovery as a function of cost and technology;

o The institutional, financial, and political considerations that might restrain or encourage production, conversion, and distribution; and

o The nonenergy materials, manufacturing capacity, manpower, capital, land, and water required for a range of energy-supply estimates.

To carry out this task, the panel organized a number of resource groups to focus on specific energy technologies and resources. To date, the reports of four of these -- on uranium resources controlled nuclear fusion, geothermal resources and technology, and solar and other renewable energy sources -- have been published separately as supporting papers of the study.*

Most of the members of the Supply and Delivery Panel and its resource groups have long been associated with companies engaged in energy supply and delivery. Those associated with universities and national laboratories have been engaged in research, development, and analysis of energy supply. The work of the panel was substantially completed by early 1977. While the panel has not attempted to update this work to 1978 or 1979, its members judge that their major conclusions remain firmly based.

The National Research Council customarily publishes only the final reports of its committees. However, the panel reports, prepared as informational documents for the committee, provide useful documentation for readers of the committee report. They are therefore also being published. The findings in these documents are those of the authors and are not endorsed by CONAES or the NRC; some of the conclusions are inevitably at variance with those of the CONAES report.

The report of the Supply and Delivery Panel was reviewed in accordance with the procedures approved by the Report Review Committee of the National Research Council. For the review, designated members of CONAES served as reviewers, rather than the customary practice of a separate report review committee consisting of members of the National Academy of Sciences, the National Academy of Engineering, and the Institute of Medicine.

---

*National Research Council. 1978. <u>Problems of U.S. Uranium Resources and Supply to the Year 2010</u>. Uranium Resource Group, Supply and Delivery Panel, Committee on Nuclear and Alternative Energy Systems. Supporting Paper 1. Washington, D.C.: National Academy of Sciences.

National Research Council. 1978. <u>Controlled Nuclear Fusion: Current Research and Potential Progress</u>. Fusion Assessment Resource Group, Supply and Delivery Panel, Committee on Nuclear and Alternative Energy Systems. Supporting Paper 3. Washington, D.C.: National Academy of Sciences.

National Research Council. 1979. <u>Geothermal Resources and Technology in the United States</u>. Geothermal Resource Group, Supply and Delivery Panel, Committee on Nuclear and Alternative Energy Systems. Supporting Paper 4. Washington, D.C.: National Academy of Sciences.

National Research Council. 1979. <u>Domestic Potential of Solar and Other Renewable Energy Sources</u>. Solar Resource Group, Supply and Delivery Panel, Committee on Nuclear and Alternative Energy Systems. Supporting Paper 6. Washington, D.C.: National Academy of Sciences.

CONAES SUPPLY AND DELIVERY PANEL

W. KENNETH DAVIS (Chairman), Vice President, Bechtel Power Corporation

FLOYD L. CULLER (Deputy Chairman), President, Electric Power Research Institute

DONALD G. ALLEN, President, Yankee Atomic Electric Company

PETER L. AUER, Professor of Mechanical and Aerospace Engineering, Cornell University

JAMES BOYD, Consultant, Washington, D.C.

HERMAN M. DIECKAMP, President, General Public Utilities

DEREK P. GREGORY, Assistant Vice President, Energy Systems Research, Institute of Gas Technology

ROBERT C. GUNNESS, Former Vice Chairman & President, Standard Oil Company of Indiana

JOHN W. LANDIS, Senior Vice President, Stone and Webster Engineering

MILTON LEVENSON, Director, Nuclear Power Division, Electric Power Research Institute

ERIC H. REICHL, President, Conoco Coal Development Company

MELVIN K. SIMMONS, Assistant Director, Solar Energy Research Institute

MORTON C. SMITH, University of California, Los Alamos Scientific Laboratory

A. C. STANOJEV, Vice President, Ebasco Services Inc.

JOHN O. BERGA, Staff Officer

HANS L. HAMESTER, Panel Consultant

SUPPLY AND DELIVERY PANEL

RESOURCE GROUPS

BREEDER REACTOR RESOURCE GROUP

Milton Levenson (Chairman), Simcha Golan (Deputy), Joseph R. Dietrich, George W. Hardigg, Herbert G. MacPherson, Roger S. Palmer, Robert H. Simon, William E. Unger.

COAL CONSERVATION RESOURCE GROUP

Eric H. Reichl (Chairman), Martin A. Elliott, J. A. Phinney, Howard Siegal.

ELECTRICITY CONVERSION RESOURCE GROUP

Herman M. Dieckamp (Chairman).

ENERGY RESOURCES RESOURCE GROUP

James Boyd (Chairman). Coal: Robert Stefanko (Chairman), William Bellano, Robert W. Erwin, Earl T. Hayes, Thomas C. Kryzer, William N. Poundstone, Michael Trbovich. Lithium: Thomas L. Kesler (Chairman), Keith Evans, Ihor Kunasz, James Vine. Oil and Gas: Charles J. Mankin (Chairman), Kenneth Crandall, Thomas M. Garland, Ralph W. Garrett, Jr., Richard J. Gonzalez, Michael T. Halbouty, Richard L. Jodry. Oil Shale: James H. Gary (Chairman), John Dew, John Hopkins, Robert McClements, Jr., Richard D. Schwendinger, John R. Donnell. Uranium: Leon T. Silver (Chairman), Jack Grynberg, Joseph B. Rosenbaum, David S. Robertson, Arnold J. Silverman.

FUSION ASSESSMENT RESOURCE GROUP

Peter L. Auer (Chairman), Keith Brueckner, Sol J. Buchsbaum, William Gough, Henry Hurwitz, Gerald Kulcinsky, Michael Lotker, Marshall N. Rosenbluth, Donald Steiner.

GAS DISTRIBUTION RESOURCE GROUP

Derek Gregory (Chairman), Henry R. Linden, Harold J. Mall, William C. McDonell.

GEOTHERMAL RESOURCE GROUP

Morton C. Smith (Chairman), James H. Barkman, Allen G. Blair, Clarence H. Bloomster, Myron H. Dorfman, Harry W. Falk, Jr., Donald F. X. Finn, Bob Greider, John H. Howard, Richard L. Jodry, J. F. Kunze, A. W. Laughlin, L. J. P. Muffler, Carel Otte, John C. Rowley, Jefferson W. Tester.

NUCLEAR CONVERTERS AND FUEL CYCLE RESOURCE GROUP

John W. Landis (Chairman), Peter W. Camp, Eugene Critoph, Andres de la Garza, Ralph W. Deuster, Albert J. Goodjohn, Robert M. Jefferson, William M. Pardue, John A. Patterson, Alfred M. Perry, Louis H. Roddis, Jr., John J. Taylor, Richard C. Vogel, Alan K. Williams, Warren F. Witzig.

OUTPUT RESOURCE GROUP

Donald G. Allen (Chairman), Bruce C. Netschert, Harry Perry, A. C. Stanojev.

PETROLEUM RESOURCE GROUP

Allen E. Bryson and Theodore Eck (Co-chairmen). <u>Domestic Refining and Facilities</u>: Allen E. Bryson, Robert C. Gunness, Ben C. Ball, Jr., Harold L. Hoffman, L. R. Roberts, David Wrigley. <u>Imports</u>: Theodore Eck, William A. Johnson, Stuart E. Watterson, Jr.

SOLAR ENERGY RESOURCE GROUP

Melvin K. Simmons (Chairman), Bruce Anderson, Piet Bos, Sheldon Butt, William Dickinson, Elmer L. Gaden, Joseph C. Grosskreutz, Ronal W. Larson, Abrahim Lavi, Aden Meinel, Marshal Merriam, Michael J. Mulcahy, Rosalie Ruegg, Roger Schmidt, Joseph L. Shay, William A. Thomas, Martin Wolf.

COMMITTEE ON NUCLEAR AND ALTERNATIVE ENERGY SYSTEMS

HARVEY BROOKS (Co-Chairman), Benjamin Peirce Professor of Technology and Public Policy, Aiken Computation Laboratory, Harvard University

EDWARD L. GINZTON (Co-Chairman), Chairman of the Board, Varian Associates

KENNETH E. BOULDING, Distinguished Professor of Economics, Institute of Behavioral Science, University of Colorado

ROBERT H. CANNON, JR., Chairman, Division of Engineering and Applied Science, California Institute of Technology

EDWARD J. GORNOWSKI, Executive Vice President, Exxon Research and Engineering Company

JOHN P. HOLDREN, Professor, Energy and Resources Program, University of California at Berkeley

HENDRIK S. HOUTHAKKER, Henry Lee Professor of Economics, Department of Economics, Harvard University

HENRY I. KOHN, Professor, Radiation Biology, Shields Warren Radiation Laboratory

STANLEY J. LEWAND, Vice President, Public Utilities Division, The Chase Manhattan Bank

LUDWIG F. LISCHER, Vice President of Engineering, Commonwealth Edison Company

JOHN C. NEESS, Professor of Zoology, University of Wisconsin

DAVID ROSE, Professor, Nuclear Engineering, Massachusetts Institute of Technology

DAVID SIVE, Attorney at Law, Winer, Neuberger and Sive

BERNARD I. SPINRAD, Professor, Nuclear Engineering, Radiation Center, Oregon State University

---

JACK M. HOLLANDER, Study Director

JOHN O. BERGA, Deputy Study Director

CONTENTS

1 INTRODUCTION	1
The Current Energy Picture	2
Energy Resources	4
Projecting Future Energy Supplies	5
The Relationship of the Supply of Energy to Its Cost	5
Supply Scenarios	6
Effects of the Business-as-Usual Approach	7
Enhanced Supply and National Commitment Energy Availability	10
Liquid and Gaseous Fuels	10
Synthetic Fuels	10
Electric Power Generation	12
Coal	13
Nuclear Power	15
Advanced Energy Sources	18
The Impacts of Alternative Approaches on Production	22
Government Incentives for Energy Supply	22
Energy Supply Contingencies	24
Lead Times in Deploying New Energy Systems	25
Summary	26
References	28
2 ELECTRICITY	29
The Nation's Electrical System	29
Electricity Demand Growth	30

Near-Term Outlook	30
Long-Term Outlook	34
Electricity Supply Scenarios	34
The Near Term	35
The Long Term	35
Fuel Mix to Meet Electricity Needs	37
Coal for Electricity Generation	44
Past and Future Uses	44
Geographical Distribution of Coal	46
Pollution Control Technology	47
Oil and Gas for Electricity Generation	47
Oil-Fired Electricity Generation	48
Gas-Fired Electricity Generation	48
Nuclear Energy for Electricity Generation	49
Present Status	49
Future Role of Nuclear Power	49
Hydroelectric Power	51
Potential New Energy Sources for the Generation of Electricity	52
Nuclear/Coal Cost Comparison	52
Capital Costs	52
Generation Costs: Fossil and Nuclear	56
Nontechnical Factors	59
Electric Utility Decision Making	60
References	65
3 OIL AND GAS	68
Petroleum	68
U.S. Domestic Production	70
World Production and Availability of Imports	72
Oil Refining	76
The Oil Delivery System	77
Secondary and Tertiary Recovery of Crude Oil	78
Oil Exploration as a Research and Development Priority	79

	Natural Gas	79
	U.S. Domestic Production	79
	Imports of Natural Gas	81
	Adequacy of the Existing Gas Delivery System	81
	Unevaluated and Unconventional Sources	83
	Synthetic Fuels	84
	Potential of Oil Shale	87
	Use of Synthetic Crudes	88
	Environmental Considerations	89
	Outlook for Liquid Fuels	89
	References	92
4	COAL	94
	Resources and Reserves	95
	Coal Production and Producibility	96
	Factors Affecting Future Coal Production	99
	Capital Requirements	99
	Manpower and Materials	101
	Transportation Requirements	101
	Institutional and Regulatory Factors	102
	Coal Use	102
	Industrial Use	103
	Feedstocks	103
	Exports	103
	Coal Conversion	104
	References	105
5	NUCLEAR ENERGY	106
	Introduction	106
	Nuclear Reactor Technology	109
	Basic Nuclear Fission Concepts	109
	Principal Reactor Types	111
	Nuclear Fuel Cycles	115
	Introduction	115
	The Uranium-Plutonium Fuel Cycle	118

The Thorium Fuel Cycle	126
Nuclear Fuel Cycle Economics	130
Nuclear Power Issues and Constraints	130
Uranium Supply	133
Nuclear Proliferation	133
Radioactive Waste Management	137
Nuclear Licensing	138
Advanced Reactor Systems	139
Practical Realities of Advanced Reactor Deployment	139
Improved Light Water Reactor Technologies	140
Advanced Converters	141
Thermal Breeder Reactors	147
Fast Breeder Reactors	147
LMFBR Development and Deployment	148
LMFBR Development Program	148
International Cooperation	150
Timing of LMFBR Deployment	150
LMFBR Economics	150
LMFBR as Complement to LWR	151
Nuclear Power Growth Scenarios	155
Business-as-Usual Scenario	156
Enhanced Supply Scenario	157
National Commitment Scenario	158
References	160
6 ADVANCED ENERGY SOURCES	163
Solar Energy	164
Direct Use of Solar Heat	165
Biomass Conversion	167
Solar Electric Technologies	168
Solar Energy Scenarios	171
Controlled Nuclear Fusion	174
Geothermal Energy	176

Hot Water Reservoirs	177
Natural Steam Reservoirs	177
Geopressured Reservoirs	178
Normal-Gradient Geothermal Heat	178
Hot Dry Rock	179
Molten Lavas and Magmas	179
Producibility	179
Advanced Energy System R&D	180
References	183
7 NONENERGY RESOURCE REQUIREMENTS	184
Scenarios	185
Low Case Scenario (L)	185
Middle Case Scenario (M)	189
High-Intensity Electric Case Scenario (HIE)	189
Supply and Delivery Panel Recommended Case Scenario (SDR)	189
Implications	189
Summary	192
Labor	193
Capital	193
Land	198
Water	198
APPENDIX: GLOSSARY OF TECHNICAL TERMS	207

1 INTRODUCTION

The ability of any energy form to meet a substantial part of the nation's needs depends not only on physical, technical, and economic considerations, but on social and political ones as well. It is necessary, therefore, to have adequate reserves and resources, along with economic and environmentally acceptable methods of extracting them, converting them to usable energy, and transporting that energy to its ultimate consumer. These basic conditions, however, are not enough. Political decisions, policies, and programs based on societal consensus can facilitate or hinder the use of an energy source no matter what its technical or economic advantages may be.

For example, the procedures of the Nuclear Regulatory Commission are especially important to the future of nuclear power. The pricing policies of the Federal Energy Regulatory Commission are vital in balancing the supply and demand for natural gas.

For coal the legal standards for sulfur oxide emissions will determine the relative demand for low-sulfur western coal and coal from the East, with its generally higher sulfur content; in fact, emission standards at some level could substantially retard the country's shift to relatively abundant coal from oil and natural gas, which are growing scarcer in relation to demand. Price controls on domestic oil and gas have encouraged their use beyond that justified by their real market value, and at the same time have discouraged development of new supplies.

Of additional concern to potential suppliers or investors is that regulatory standards be consistent and predictable, as well as providing benefits consistent with their costs. Consistency and predictability are of prime importance, even if it means that the restrictions are initially more stringent than industry and investors would like them.

These and similar considerations will determine what combination of energy sources will be most economical, efficient, and politically acceptable in the future, and thus what will be available.

THE CURRENT ENERGY PICTURE

During the several decades preceding the 1973-74 oil embargo, energy was produced in the United States in increasing quantities and at decreasing prices relative to the average price of all goods and services. This trend resulted from continually improving technologies, rising productivity, growing production and distribution capacity contributing to economies of scale, and the discovery of large sources of oil and gas in several regions of the world. Since the early 1970s, however, the picture has been quite different. Most large and easily accessible U.S. oil and gas fields have probably been discovered, existing domestic resources are being depleted, and the United States relies more and more each year on relatively expensive imported oil. Combined with the costs of increasingly stringent environmental protection requirements, these factors act relentlessly to increase the real prices of all forms of energy.

Table 1 shows the pattern of mineral resource consumption and availability (estimated ultimately recoverable resources) for the United States. It is evident that the present reliance on oil and natural gas (74 percent of energy consumption) is not sustainable, since these resources represent only about 1 percent of economically recoverable resources.

Specifics of the current U.S. energy outlook include a number of variable factors beginning with: increasing demand for all forms of energy, strongly linked to economic activity and thus to employment; declining domestic production of oil (except for a temporary increase due to Alaskan production) concomitant with growing demand for it; declining domestic production of natural gas and natural gas liquids from both land and off-shore areas; rising imports of oil and oil products, mostly from OPEC and increasingly from members of the Organization of Arabian Petroleum Exporting Countries (OAPEC); and finally, increasing competition for worldwide oil supplies from both industrialized and developing countries, leading to increasing international political pressure on the United States to reduce imports.

Coal and uranium are the only major domestic energy resources that will be available for significant expansion for the next several decades. However, both the coal and nuclear industries are hampered by regulation and political factors. One serious development is that between 1975 and 1978 new orders for nuclear plants numbered only six; cancellations of existing orders far exceeded this number, resulting in a net decrease in utility commitments. Although it less well recognized, net orders for coal-fired capacity have declined as well, due to a large number of cancellations. Coal production is at less than capacity because the market for coal is inhibited by environmental considerations and uncertainty in the industry, and there is little incentive to expand the capacity of mining or transportation facilities until it is clear that demand will increase substantially. Future prospects, therefore look bleak unless federal policies and practices change. Potential increases in hydroelectric generation are limited by the scarcity of economic and environmentally acceptable sites, although there may be some opportunity for small, low-head hydroelectric installations.

Table 1 U.S. energy consumption and resources in 1977, by percentages of total

Energy source	Current consumption	Relative abundance of energy resources	
		Without breeder	With breeder
Oil (including natural gas liquids)	49	0.5	0.3
Natural gas	26	0.5	0.3
Coal	19	83.9	44.7
Nuclear power (uranium)	3	1.1	47.3[a]
Oil shale	0	14.0	7.4
Nonmineral energy sources	3	--	--
Total	100.0	100.0	100.0

[a]This estimate is conservative, since it is based only on uranium resources economically recoverable for use in light water reactors. Breeder reactors, however, would be able to use very low grade ores, which tends to magnify the potential resource base.

Although there are major efforts under way to develop several advanced energy sources, including the various solar technologies and nuclear fusion, they are not yet developed to the point at which they could be economic, and they cannot be counted on as large, reliable energy supplies during the rest of the twentieth century.

Lead times for licensing, designing, and constructing energy supply facilities of all types are now as much as 10 to 12 years for established technologies. This makes effective planning difficult and decreases the probability of meeting future energy demands. For example, the planning horizon recognized by most state public utility commissions is only 10 years. The result is that investment in new energy supplies is considered by many an unacceptable risk at this time, principally because of uncertainties in government policy, regulatory requirements, and innumerable delays. Major new technologies can take 20 to 30 years or more to reach widespread use, and therefore will not be major factors until at least the end of the century.

ENERGY RESOURCES

As the preliminary step in developing a strategy for decreasing the dependence on oil and gas and shifting the pattern for consumption toward our more abundant domestic resources, the Supply and Delivery Panel assessed estimates of the quantities of recoverable reserves and resources of the principal energy minerals found in the United States and adopted those that appeared most reliable.

Table 2 presents the energy contents of these reserves and resources, in quadrillions of Btu, or "quads." (The substantial additional energy potentially available from geothermal, solar, and nuclear fusion sources are not included.)

Table 2 U.S. recoverable reserves and resources of energy minerals, in quads

Mineral	Recoverable reserves	Ultimately recoverable resources[a]
Coal	6,034	79,365
Oil (including natural gas liquids)	201	466
Natural gas	233	495
Oil shale	0	13,200
Uranium (light water reactors, once-through fuel cycle)	247	680
Uranium (breeder)[b]	25,000	84,000

[a] Resources include all deposits known or believed to exist in such form that economic extraction is currently or potentially feasible. Recoverable reserves are those specifically identified resources that can be extracted economically with existing technology at prevailing prices.

[b] Assumes recovery of 70-percent of the energy content of natural uranium, plus 12,000 quads available from depleted uranium tails stockpile.

Source: Supply and Delivery Panel resource group reports (available in CONAES public file).

An examination of the relative abundance of these resources compared to the current pattern of energy use makes several points obvious. First, annual domestic consumption of energy (76 quads) is only a small fraction of recoverable reserves. Nevertheless, present annual oil and gas consumption (37 and 20 quads, respectively, together totaling 75 percent of total energy consumption) are alarmingly high compared to the estimates of recoverable reserves and resources. Coal reserves and resources, however, are large compared to the 1977 consumption rate of 14 quads. Oil shale is a potential source of large supplies of hydrocarbon liquids in the long run, although its contribution in this century will be rather small. Breeder reactors can increase energy extraction from uranium resources many times over (Table 2). With only light water reactors, however, the uranium resource is comparable to that of oil or gas. Geothermal, solar, and nuclear (fusion and breeder) resources are virtually inexhaustible, and their development should, therefore, be assigned a high priority.

PROJECTING FUTURE ENERGY SUPPLIES

The Relationship of the Supply of Energy to Its Cost

In estimating the producibility of energy in the future, the panel recognized the desirability of developing a relationship between supply and price. Under ideal free-market conditions, of course, the demand for supply of a given energy source would be determined relative to the prices of alternatives, as well as the rate with which alternatives can be substituted. An assumption about future trends in the prices and production costs of the various energy products would allot theoretical estimates of the market share for each product at any time.

In its early meetings, the panel considered the possibility of developing such estimates. In the judgment of the panel, however, prices, costs, and consumption of energy sources are so greatly affected by regulation and other expressions of national policy that this approach was abandoned. The availability of an energy source is much more likely to be determined by regulatory conditions and their impacts on investment than by prices or production cost. Accordingly, the panel chose to base its estimates on a range of hypothetical "scenarios," based on sets of assumptions about these conditions. Each resource group was charged with investigating the prospects for producing its assigned energy source under "business-as-usual," "enhanced supply," and "national commitment" conditions. The specific conditions assumed varied from resource group to resource group, depending on the particular characteristics of the individual energy source. The definition of these sets of conditions necessarily included certain assumptions about price or cost, such as whether prices were controlled, governmental subsidies or tax incentives provided, environmental regulations increased or decreased, and so on. In no case, however, could the panel develop a unique set of curves of supply producibility versus price (or cost) that would be generally applicable. In any case there would be wide variation from one location to another in the United States, depending on the characteristics of each fuel (sulfur content, crude oil gravity and type, heating value, transportation costs, and so on), as well as changes with time that would themselves depend on previous history. It is extremely important to recognize that few choices are made solely on the basis of current prices; individual expectations of future prices and conditions of use strongly influence such decisions.

While the Supply and Delivery Panel was not responsible for estimates of demand--that being the responsibility of the Demand and Conservation Panel--it needed nonetheless to develop a demand basis for its own use. The Supply and Delivery Panel believes that energy demand in the United States--as well as in the rest of the world--will continue to grow, albeit at a reduced rate, until well beyond the 2010 horizon of this study. The panel is generally less optimistic than some others involved in the CONAES study about the extent to which growth in demand will be affected by conservation.

SUPPLY SCENARIOS

The three alternative supply scenarios for each energy source differ mainly in the assumed degree of commitment to developing and deploying additional energy supply. These scenarios, whose specific provisions are described in succeeding chapters, can be summarized in general terms as follows:

- Business as Usual. This scenario is based on the assumption that existing attitudes, policies, and practices are extended into the future with little change, and that integrated, effective energy supply policies are not established and implemented--in short, that the steps necessary to insure adequate energy supplies are not taken by national policy makers.

- Enhanced Supply. This scenario depends on the assumption that a well-balanced, comprehensive set of energy supply policies is enacted and aggressively pursued, that decision making and regulatory actions are timely and coordinated, and that promising new technologies are appropriately supported.

- National Commitment. This scenario assumes that the same comprehensive set of energy policies is pursued as in the enhanced supply scenario, but more aggressively in specific areas. Energy supplies are given the highest priority in allocating national resources, and calculated risks are taken in deploying promising new energy technologies before they are technically proven and economically competitive.

Although estimates of the energy production potential from each source were made for each of the scenarios, it is improbable that all sources could be accorded special emphasis at any one time. For instance, it is not likely that enhanced supply conditions could be achieved simultaneously for all technologies, and extremely unlikely that a national commitment could be supported for more than two or three sources at once. It is also important to note that the scenarios described presume technical and economic viability; some prospective technologies now considered promising may not be successfully developed no matter what the degree of commitment. Even a technically successful development program may result in an unacceptably costly technology.

Obviously, scenario estimates must be used with caution. One should not conclude that a variety of options from which to choose is available, or that there is adequate support for all of the proposed research and development programs. This is far from being true. In the opinion of the panel, the picture that emerges from the Supply and Delivery Panel analyses indicates a need for much more aggressive national actions, both with respect to research and development for the future and implementation of existing options. The considerations fall logically into two time periods, one from now until about 1990, which is the short term, and the second from about 1990 to 2010, which we see as the intermediate range. The difference between the two is that we have little alternative but to use the options already available to us for the period into the early 1990s as new developments are not likely to make any significant impact before that time. The development of new energy options (only future possibilities today) may lead to new solutions in the 1990 to 2010 period and in the longer range beyond the year 2010.

Effects of the Business-as-Usual Approach

The panel examined the implications for energy supply if present policies and practices continue. The business-as-usual scenario does not imply that changes will not occur—only that such changes will not be subject to a coordinated policy that recognizes the critical nature of the energy situation. This approach also implies a continuance of the complacency prevalent in this country.

Under business-as-usual assumptions, price controls on domestic production of oil and gas would remain in effect at levels well below world market prices. The future of environmental protection requirements would remain subject to unpredictable and sudden change, discouraging investments in new production. Delays in leasing and withdrawal of public lands from energy development would occur in two stages, with separate permits for exploration and production.

Under these conditions, domestic production of both oil and gas would continue to decline (Table 3), though the Alaskan North Slope discovery will increase production for several years. The growing gap between domestic supply and demand is now offset by increasing hydrocarbon imports, but by the 1990s growth in world demand and leveling off in world production are likely to make such a policy infeasible. Although the shortage of hydrocarbon fuels could in principle be relieved by substituting fuels such as shale oil and synthetic oil and gas from coal, these technologies are not now economic and face technical, institutional, and environmental problems. Since there would be no incentive for timely large scale development and deployment of synthetic fuels under business-as-usual conditions, these fuels would become available much too late to reduce significantly the demand for imports. In this scenario synthetic fuel production might also be constrained after 1990 by insufficient supplies of coal.

Table 3 Domestic energy production for business-as-usual scenario, in quads per year

Energy source	1977[a]	1990	2000	2010
Crude oil (including natural gas liquids)	19.6	16.0	12.0	6.0
Natural gas	19.4	10.3	7.0	5.0
Shale oil	0	0	0	0
Synthetic liquids[b]	0	(0.3)	(2.3)	(6.1)
Synthetic gas[b]	0	(1.3)	(3.5)	(4.1)
Coal	16.4	25.0	34.0	42.0
Geothermal	0	0.4	0.9	2.4
Solar	0	0	0.1	0.3
Nuclear	2.7	10.0	12.5	15.8
Hydroelectric	2.4	4.0	5.0	5.0
Total	60.5	65.7	71.5	76.5

[a] 1977 data from U.S. Department of Energy (1978).

[b] Synthetic fuels are produced from coal and oil shale, and therefore are not added in the totals.

Under business-as-usual assumptions, coal production could expand under pressure of the need to replace oil and gas in many applications. However, as we shall see, problems associated with leasing of federal lands and the lack of consistent environmental policies and regulations, and the need for upgraded transport facilities, leading to cost increases and delays, may limit production below what is demanded. Although coal is needed for many purposes and its vastly increased use is prescribed as a means of lessening the demand for oil and gas, the uncertainties associated with its production and delivery will definitely limit growth given current policies.

Business-as-usual conditions would even more drastically limit the growth of nuclear power. Uncertainties due to a continued lack of positive and consistent government policies, organized antinuclear opposition, increasingly complex regulatory and environmental requirements, and continuing delay in dealing with the waste disposal and reprocessing issues, has caused utilities to virtually stop ordering new nuclear generating capacity. Indeed, the nuclear industry may not survive if orders do not resume soon. Even if ordering resumes at a satisfactory pace, continued deferral of a breeder reactor technology demonstration and spent fuel reprocessing and recycling, as well as the failure to demonstrate a radioactive waste disposal method, could well result in a gradual phasing out of nuclear construction starting around 1990-2000. At about the same time, anticipation of uranium fuel shortages within the lifetimes of reactors built during that period would contribute to this cessation of activity. Given these conditions the United States will find itself unable to meet the demand for electricity at that time without encountering the serious environmental and public health risks of using coal-fired units for almost all new generating capacity.

Hydroelectric power generation capacity can grow only slowly, and its ultimate contribution will be small, because the best sites are already in use and environmental restrictions on the remaining sites are likely to limit the number of new hydroelectric stations. Its contribution as a percentage of electricity generation is expected to decline.

Advanced technologies such as geothermal heating and solar heating and cooling might in the near future make some contributions, but in the business-as-usual scenario these would be negligible. Similarly, solar electric technologies, more advanced applications of geothermal energy, and other advanced electric power technologies could make relatively minor contributions. Fusion power plants, if proven feasible, are not expected to contribute significantly under any circumstances until well after 2010.

Table 3 summarizes what might be produced under business-as-usual conditions. It does not include oil imports. Without imports, only about 16 quads of domestic liquid fuels may be available annually in 1990, and only about 12 in 2010. The United States now consumes more than 35 quads of liquid fuels per year, about 20 quads of which come from domestic oil production and the rest from imports (at a cost of about $2.5 billion per quad). Obviously, under business-as-usual conditions we would have either to reduce our consumption drastically or rely increasingly on imports (which is likely to be impossible), or both. All of this suggests that additional measures to foster development and production of energy resources are urgently required.

ENHANCED SUPPLY AND NATIONAL COMMITMENT ENERGY AVAILABILITY

The shift toward more abundant resources will take time, require large capital investments, and involve great numbers of people and quantities of materials. The substantial financial risk involved is not likely to be accepted by the private sector unless clear, positive signals and incentives are provided by the federal government.

Liquid and Gaseous Fuels

The demand for hydrocarbon liquids and gases for electricity generation can ultimately be reduced to very low levels by depending on coal-fired and nuclear units for all future baseload capacity additions. Only about 10 percent of the oil used in this country is burned in power plants, however, and about 15 percent of the gas. Thus, even if the current policy of curtailing the burning of oil and gas in power stations is successful, the effect on demand for these fuels will be relatively small, and similar reductions will need to be realized for transportation, space heating, chemical feedstocks, and industrial uses. However, even if conservation efforts realize significant energy savings in these areas, total demand for these fuels seems certain to continue to exceed domestic production by a substantial and growing margin.

Domestic production of petroleum and natural gas liquids is expected to decline in the future regardless of the measures taken to augment it. Shown in Tables 4 and 5 are the Supply and Delivery Panel estimates for future production under the three scenarios. The assumptions of the enhanced supply scenario include accelerated federal offshore leasing, decontrolled wellhead oil and gas prices, evolutionary improvements in exploration and production technology, and streamlined permit processes. The national commitment scenario depends on relaxation of some Clean Air Act stipulations, streamlined environmental impact statement procedures, federal support for the development and application of new technologies, federal return of previously withdrawn lands for development, and government priorities on materials and labor for oil development. Appropriate government actions would significantly retard the decline expected under a business-as-usual scenario, but the difference in production between enhanced supply and national commitment cases is not great. The assumptions for both the enhanced supply and national commitment scenarios for gas are as described for crude oil.

Synthetic Fuels

Petroleum and natural gas are not likely to be produced domestically in sufficient quantities under any scenario. The only expected major new domestic sources of substitutes are oil from shale, and oil and gas from coal, neither of which is included in Tables 4 or 5. It is estimated that shale oil may not be produced at all if present trends continue, but that with enhanced supply conditions it could be 1.5 quads per year by 2010, and with a national commitment about 3 quads per year.

Table 4 Oil and natural gas liquids production potential, in quads per year

Scenario	1975	1985	1990	2000	2010
Business as usual	20	18	16	12	6
Enhanced supply	20	21	20	18	16
National commitment	20	21	21	20	18

Table 5 Potential natural gas production, in quads per year

Scenario	1975	1985	1990	2000	2010
Business as usual	19.7	13.5	10.3	7.0	5.0
Enhanced supply	19.7	16.1	15.8	15.0	14.0
National commitment	19.7	18.5	18.0	17.0	16.0

If a national commitment is made, the contribution from coal-based synthetics could be about 12.9 quads per year of synthetic liquids, and 7.9 quads per year of synthetic gas by 2010. (Again, not included in Tables 4 or 5.) These values are based on the assumption that liquid production is emphasized over gas production after 1995. To reach these levels for coal-based synthetic fuels, immediate action is required. The aforementioned values are based on the assumption that government authority is used to underwrite product prices; expedite construction and operating permits; and allocate labor, money, and materials on a priority basis as necessary.

For some time into the future, synthetic fuels will be more costly than crude oil and natural gas from conventional sources. Unless some form of incentive is accorded producers, their development is likely to be delayed until the costs of conventional and synthetic fuels are closer to equal. After that, it would take time for the new technologies to be developed and deployed commercially.

If, then, synthetic fuels are to contribute to domestic energy supplies by the time they are required, provisions must be made now for incentives to large-scale deployment of these technologies. One proposal for providing such incentives is to require that a certain fraction (increasing with time) of hydrocarbon products be of synthetic origin. In any case, some form of guaranteed market or other financial support would be needed to encourage their early development.

Therefore, as synthetics are introduced relatively slowly, there will be a dampening effect on fuel price increases. Synthetic fuels are inherently less susceptible to inflationary price increases than natural fluids because most of their cost is associated with the capital costs of plants.

Vital to the introduction of a synthetic based on coal or shale is the provision of adequate water supplies to meet process needs. While numerous studies aimed at quantifying these water requirements have been performed, specific plans for water supply and use must be developed early for the western states, where such fuels could be produced most economically. If the higher levels of synthetic production are expected, much more water-conservative processes or interbasin transfers of water may be needed.

Electric Power Generation

Electricity is a versatile, convenient, and precisely controllable form of energy. For these reasons its rate of growth has generally exceeded that of total energy consumption. This trend has persisted despite the transition in the last few years from decreasing real prices for electricity. Electricity is generated today from coal, oil, gas, nuclear, hydroelectric, and geothermal sources. Of these, only coal and nuclear power offer the promise of greatly expanded use. Coal provided about 46 percent of the nation's electricity; oil, gas, and hydroelectric, 42 percent; nuclear, 12 percent; and other sources (geothermal, wood, and waste), a small fraction of 1 percent. As noted earlier, oil and gas are in declining production, and their use in electricity generation is expected to fall accordingly. These fuels are not essential for electricity generation but they are vital to other uses, such as transportation. Obviously a high priority must be given to substituting coal and nuclear power for oil and gas in electricity generation.

Utilities are already moving away from oil-fired generation because of the high price of imported oil, uncertainties about its future availability, and the pressure of government regulations. This is not to say that oil use by utilities is declining; only that less new oil-fired generating capacity is being planned. Because of the long lead times for new plants, oil consumption by electric utilities is expected to continue rising until it levels off about 1982 and begins a gradual decline. Conversion of oil-fired plants to coal, which is generally impracticable, will have only a small impact in the near future.

Burning of natural gas in power plants peaked in 1971, and is expected to decline steadily to negligible amounts by 2000. The National Electric Reliability Council (1977) projects that by 1985 natural gas use in electric utilities will be only 30 percent of the 1971 peak.

It seems clear that coal and uranium must become the predominant fuel sources for electricity generation and are likely to retain this position until the advanced technologies begin to make major contributions, some time after the turn of the century. Both coal and nuclear sources together will be needed even if electricity demand growth is very low. Even if either alone were sufficient, diversification of supply would be desirable to promote competition and minimize the potential for energy shortages in the event of a major interruption of a single fuel supply. Just how fast the use of electricity will increase is, of course, very uncertain. There have been major changes in the growth rate during the last few years, and forecasts vary widely. At the upper extreme is the historical growth rate of 6 to 7 percent per year, which prevailed from the mid-1950s to the early 1970s. The panel believes that growth in the future will probably be lower than this, due to the combined impact of lower economic growth, higher electricity prices, and conservation efforts. On the low side, if business-as-usual conditions continue for all electricity generation sources, it is estimated that electric capacity might only double by the year 2010--a growth rate of 2 percent per year--although the panel believes the actual growth in demand will be greater than that, due to the need to substitute electric power for hydrocarbon fuels, the electricity demands of pollution control devices, and similar factors. Based on this range of estimates, the Supply and Delivery Panel suggests that an average annual growth rate of 4 to 5 percent should be used as a guideline for prudent planning; this would amount to an approximate quadrupling of the nation's generating capacity.

Coal

Economically recoverable coal reserves are very large, having an energy potential of about 6000 quads. For most uses, however, it is not a preferred fuel in its natural form, so that most proposals for expanding its use call for coal to be converted to liquids, gases, or electricity, highly preferable forms of energy. Indeed, it is this capacity for conversion to end-use fuels that enhances its future growth potential.

In the near term, coal production is likely to be limited by demand, not by the production capacity of the industry. The current availability of more desirable fuels and questions about the environmental acceptability of coal combustion make future demand highly uncertain. The projected increase in production is based on the assumption that more consistent policies will be established and maintained to provide some stable basis for planning.

Table 6 lists the panel's estimates of coal production potential under each of the three sets of supply scenario conditions. The wide difference between the enhanced supply and national commitment scenario's assumed increases in coal mine productivity and capital availability, as well as streamlined regulatory requirements.

Table 6 Projected coal production, in quads per year

Scenario	1976	1990	2000	2010
Business as usual	13.3	25.0	34.0	42.0
Enhanced supply	13.3	26.6	37.2	49.5
National commitment	13.3	32.5	75.0	100.0

If, as the panel recommends, a national commitment is made to the production of synthetic fuels from coal, about 21 quads of synthetic gases and liquids could be produced annually by the year 2010. This would be at the expense of about 31 quads of coal, and at enhanced supply coal production levels it would likely be difficult to provide coal for all other uses.

Whether the levels for coal production of the enhanced supply scenario will be achieved depends on the resolution of a host of problems now facing the industry, including the need for clear and unified federal energy policies, particularly on leasing federal lands containing coal; establishment of stable environmental regulations on mining and use of coal; formulation and implementation of a detailed and comprehensive water management and supply plan for the western states; acceleration of impact statement review and advisory hearing and permitting processes; reduction of uncertainties on limits for the release of carbon dioxide in the atmosphere; and promotion of policies to make capital funds more easily available to coal mining and coal conversion industries. Other problems include the need to ensure that enough coal miners and mining engineers will be available when needed, that equipment-intensive methods will be further developed to minimize labor requirements, and that the coal transport system will be continuously upgraded. Emphatic policies and funding could solve most of these problems; the principal exception is the potential effect of rising atmospheric carbon dioxide levels on climate, which requires further study. Possible problems associated with CO_2, however, are of a long-term nature and should not constrain coal for the next few decades.

Nuclear Power

As noted earlier, only coal and nuclear power can be domestically available in the next few decades in sufficient amounts to meet electricity demand. Because other demands for the use of coal are numerous, the supply of nuclear power is, therefore, vital for electricity generation during the 1990-2010 period.

As of February 1979 domestic nuclear power capacity consists of 72 reactors in operation, contributing about 10 percent of electrical capacity and about 12 percent of actual generation. Plants under construction or on order will bring the total nuclear generating capacity to 200 GWe (about 11.7 quads per year) by 1990. Major uncertainties are discouraging additional orders so much that effective new government policies will be required if nuclear energy is to continue to grow past about 1990.

Given success in implementing such policies, the future contribution from nuclear power depends on the total growth rate of electricity, the producibility of nuclear fuels, and the technology used to extract energy from the fuels. The existing U.S. nuclear electricity industry is based on the light water reactor (LWR), which uses uranium fuel relatively inefficiently, particularly as operated today, without recycling uranium and plutoniom recovered from spent fuel. The energy recoverable from estimated U.S. uranium resources by LWRs is about the same as that recoverable from oil or gas. Nuclear power using LWRs only would reach its peak shortly after the end of this century unless its growth were slow or the amount of uranium discovered and produced is significantly greater than now estimated.

A number of other reactor designs could potentially provide more efficient use of fuel. Such reactors generally are classed either as advanced converters or as breeders, depending on the conversion ratio (fissile atoms produced to fissile atoms destroyed during the fission process). Compared to present LWR conversion ratio of about 0.55, advanced converters achieve ratios in the range of 0.7 to 1.0 and breeders achieve ratios greater than 1.0.

The higher conversion ratios of advanced converters offer a marginal improvement in the use of uranium. They are still, however, net consumers of fissile fuel, so that their contribution is limited by the size of the resource base. On the other hand, breeder reactors produce more fissile material than they consume, and their widespread use would provide the fissile fuel to support long-term growth in nuclear power.

The Uranium Research Group of the panel (National Research Council, 1978) assessed the availability of uranium resources, using Department of Energy estimates based on confidential and proprietary industry information. The resource group's estimates were then modified by information received by the panel from industry, other government agencies, and academic sources (Energy Research and Development Administration, 1976a, 1976b, 1976c). In their judgment, the best estimates of a combination of economically recoverable reserves and potential resources of uranium in the United States are 1.76 million tons of U_3O_8 at a forward cost of thirty dollars per pound. They also noted that U.S.

and world reserves of thorium (a potential fuel for advanced converter reactors) have never been adequately appraised and that no appraisals are underway. On this basis, the panel believes that for purposes of planning we should assume this level of uranium ore--a level that would provide lifetime fuel supplies for about 500 gigawatts of LWR capacity even if spent fuel is reprocessed so that its uranium and plutonium can be recycled. The panel also recommends a major commitment to uranium exploration if nuclear power is to be a major contributor to total U.S. energy supply.

If electricity growth is moderate or high, uranium resources are likely to limit the contribution of nuclear power until either advanced converters or breeders are introduced. The lead time for development, demonstration, and introduction of any new nuclear reactor type is expected to be 15 to 20 years. As noted previously, advanced converters are net consumers of fissile fuel because of the heavy initial fuel loadings. It is because of this that they provide no real advantages until after at least 10 years in operation. It is, therefore, too late for advanced converters to contribute to a major savings in the resource base--unless growth is very low, in which case there would be no market to support the introduction of any new technologies.

Bearing this in mind, the panel urges that the development and demonstration of breeder reactor technology proceed without delay, so that breeders will be available when and if needed to support continuing growth in nuclear electricity. Since the technology for the liquid metal fast breeder reactor (LMFBR) is already well advanced, and many years ahead of other concepts, the panel recommends that its development be expedited. Timely development of LMFBR technology will require a concerted national effort. In case LMFBR development runs into unexpected difficulties, funding for a backup program to develop the gas-cooled fast breeder reactor parallel with the LMFBR would be prudent.

The fundamental question is: where do we go from here? The progression of reactor development from enriched uranium light water reactors to plutonium-fueled breeder reactors has always been the essence of the long range nuclear power strategy in most of the technically advanced world. The only real questions were the precise nature and characteristics of the light water and breeder reactors, and the timing for beginning the switch to breeders. The key influence on this transition was uranium cost--for as uranium is consumed and its price increases, the incentive to switch to fast breeders becomes greater. Other factors that would influence this timing include industrial production capability and relative capital cost.

Furthermore, as a technology traverses the gap between scientific feasibility and commercial viability, the ever increasing cost of research and development will have a strong impact on the scope and pace of future energy research and development. The choice of particular technologies for development and subsequent deployment will continue to be influenced by political considerations as well as by technical factors such that intense competition will persist for limited federal funds. Accordingly, future nuclear research and development should focus on those technologies that have a high probability of commercial success as well as the potential of supplying substantial generating capacity.

It should be recognized that this intensive effort to develop and demonstrate breeder technology does not imply a corresponding commitment to deploying breeders. If, however, the quantity of uranium resources proves as limited as the Uranium Resource Group suggests, a shift to breeder reactors on a national commitment basis is likely be be necessary to meet the long-term demand for electricity. A rapid deployment program (taking full advantage of overseas technologies) may have to be launched within the next several years if the potential uranium supply outlook does not improve by that time. If additional uranium is discovered, the decision to deploy breeders can be deferred.

Potential Nuclear Generating Capacity

More than that of any other source, the potential contribution from nuclear power depends on the national policies adapted. In the event that existing practices and trends continue without appreciable change (business as usual), the nuclear outlook is bleak. If this is so, the panel estimates that the existing orders will be filled on a delayed schedule, new orders will be limited, and the nuclear industry manufacturing capacity will decline substantially (Table 7).

If it is assumed that enhanced supply policies are adopted, to reduce many of the uncertainties surrounding nuclear power, much higher levels of nuclear capacity are possible (Table 7). In this case, the estimates are based on the assumption that there are strong government commitments to closing the LWR fuel cycle and to building LWR plants at the full industry capacity of 25 to 30 units per year.

Finally, the panel estimated the potential contribution under the assumption that even stronger policies were enacted, including a full national commitment to the earliest possible development and introduction of LMFBRs. This scenario obviously assumes not only successful development of LMFBR technology, but also the need for wide and early deployment, although the need for such rapid deployment is not yet apparent.

With the renewed emphasis on uranium the financial climate for exploration in the future will be determined largely by the extent to which government policy on nuclear power is perceived by the uranium industry to assure a long-term market for uranium; tax incentives; the changing structure of the energy industry; extent of environmental constraints on exploration, mining, and milling; and policies on access to public lands. With an appropriate atmosphere, the electric power industry, as well as the petroleum and mining companies, could develop the required capital. Improving uranium exploration technology requires research and development, especially in the basic sciences.

Nuclear power is not alone with respect to the multitude of institutional problems inhibiting deployment of power generation facilities. Utility efforts to license and construct coal-fired installations as well as hydroelectric projects are also fraught with obstacles, and it is not unlikely that advanced energy technologies will also be faced with efforts to block their demonstration and deployment.

Table 7 Installed nuclear capacity, in gigawatts electric

Scenario	Reactor type	1990	2000	2010
Business as usual				
	LWRs	165	210	260
	LMFBRs	0	0	0
Enhanced supply				
	LWRs	220	500	700
	LMFBRs	0	2	10
National commitment				
	LWRs	240	540	750
	LMFBRs	0	10	100

Advanced Energy Sources

Energy sources expected to supply most domestic energy needs for the rest of this century are derived from exhaustible resources. As these sources continue to be depleted, there is concern about their availability at acceptable costs, thus providing a strong impetus for development of alternative sources.

Inexhaustible energy sources discussed in this report include solar energy (derived both directly and indirectly from sunlight), thermonuclear fusion, and geothermal energy. Along with nuclear fission using breeders, these energy sources constitute all the energy sources now known that have the potential for providing a major fraction of total energy needs.

Appraising the potential contribution of these resource areas to the future supply of energy is difficult, since none of the technologies is fully developed and each is at a different stage of deployment. For technologies that are fairly well developed, such as breeders or solar heating, economic competition and the rate of market penetration must be estimated. For those relatively undeveloped, such as fusion or hot rock geothermal, the probability that technical viability and competitive cost will be achieved, as well as the rate of market penetration, must be evaluated. Thus, the magnitude and timing of contributions from these sources should not be overstated. While the panel strongly supports an extensive development and demonstration program to determine which technologies will emerge as economically viable energy supplies, it advises against reliance on their promises.

Solar Energy

The principal means of using solar radiation include technologies for heating and cooling buildings, industrial process heat, bioconversion, solar thermal conversion, photovoltaic conversion, wind energy conversion, and ocean thermal energy conversion. Interest in these various technologies has increased dramatically in recent years.

Solar energy is claimed to offer major advantages, including generally benign environmental impacts, and, of course, inexhaustibility. In addition, it is favored by widespread public acceptance and relatively uniform geographical distribution of the resource. Certain solar technologies are suitable for use in small units, and many are considered potentially less complex and hazardous than other energy supply technologies now in use.

On the negative side, solar energy requires very large collectors to capture the incident radiation. Solar radiation is intermittent, subject to interruptions not only on the daily solar cycle but also during cloudy periods. Most solar technologies make heavy use of nonrenewable resources in their construction, and this results in high energy costs.

Despite the inability of solar technologies to compete in today's energy market (except possibly in space heating and domestic water heating), the Supply and Delivery Panel supports continued strong research and development efforts. Research should be directed at reducing costs of presently available technologies by improved techniques for collecting, storing, and distributing solar energy, but also by seeking new configurations, materials, and processes for converting solar energy, especially for the generation of electricity and of portable fuels.

Solar technology for domestic water heating is advanced enough to be used now. Solar heating of new buildings is also likely to be employed to a considerable extent in areas where the climate is favorable. Solar air conditioning, however, is particularly expensive and will require major breakthroughs before it can compete with conventional alternatives.

Solar process heat for industrial and agricultural purposes offers considerable potential, but the reliability and economics of these systems require demonstration before any significant deployment can be expected. Certain industrial applications of solar heating are particularly attractive because their year-round demand for energy enables a high utilization of the capital investment as compared with cyclic applications such as space heating.

The quantity of energy likely to be recovered by the various biomass conversion processes is not precisely known. The panel believes the quantity of such potential fuels likely to be available to be considerably less than recent optimistic forecasts.

Major research and development programs in centralized solar thermal generation of electricity are underway. The development of total solar systems combining heating, cooling, and electricity production in buildings is also underway. It is unlikely that such systems will be widely used until costs, including the cost of the necessary energy storage capacity, are substantially reduced. The present demonstration program does not advance the goal of cost reduction, in the opinion of the panel.

Solar photovoltaic conversion merits a high priority development program, but requires a fundamental breakthrough in collector technology to reduce costs. Here again, the existing demonstration program does not appear likely to lead to economically competitive systems. Like solar thermal conversion, photovoltaic systems require the development of economic energy storage systems.

The potential contribution of wind energy conversion is limited by the availability of suitable sites, the intermittent and unpredictable nature of the weather, and costs that are not generally competitive, except in special situations.

The future of ocean thermal energy conversion--generation of electric power by the thermal energy in tropical oceans--is also most uncertain. Estimates of capital costs vary widely. The low temperature differentials available require very large, and hence costly, heat exchange surfaces. Biofouling of these surfaces may be a serious operational problem. Field experiments conducted over the next several years should aid in determining whether or not further development of this concept is warranted.

The potential contributions of solar energy in all forms as estimated in the panel's scenarios is presented in Table 8. The uncertainty in these estimates is considered greater than that in the estimates of energy production presented for the more established technologies, since their technical and economic success is much less predictable.

Table 8 Solar energy production potential, in quads per year

Scenario	1985	1990	2000	2010
Business as usual	0.0	0.0	0.1	0.6
Enhanced supply	0.9	1.7	5.9	10.7
National commitment	1.6	3.3	13.1	28.8

Geothermal Energy

The heat contained in the earth's interior represents an enormous reservoir of thermal energy. So far geothermal energy can be extracted only where naturally circulating groundwater has brought it to, or nearly to, the earth's surface as steam or hot water. Geothermal energy exists in other forms, however, and work is proceeding on means of exploiting the heat in dry rock and even in molten rock. The total resource is so vast that almost any uncertainty factor could be assigned without altering the conclusion that, in the foreseeable future, commercial production of geothermal heat will not be limited by the amount of the accessible energy supply.

The principal constraints on rapid commercial development of geothermal power illustrate the difficulty of the transition from potential to reality. Geothermal energy systems are capital-intensive, the period between initial investment and initial return is long, and the rate of return to potential investors is uncertain. Nevertheless, because the energy resource is so large it warrants concentrated work. Table 9 lists this panel's scenario projections of the potential contributions of this energy source. The panel recommends that development, demonstration, and deployment of geothermal technology proceed on an enhanced supply basis.

Table 9 Geothermal production potential, in quads per year

Scenario	1980	1990	2000	2010
Business as usual	0.1	0.4	0.9	2.5
Enhanced supply	0.1	0.6	1.6	4.1
National commitment	0.2	2.2	7.8	19.9

Controlled Nuclear Fusion

Fusion is the combination of two lightweight atomic nuclei, such as deuterium and tritium, into a single nucleus of heavier mass, such as helium, with the resultant loss of the combined mass being converted into energy. To be successful, fusion technology must evolve through three stages: scientific feasibility, engineering feasibility, and commercial feasibility. Although considerable progress has been made, scientific feasibility—that is, a controlled reaction, in a laboratory, in which the energy coming out of the plasma equals or exceeds the energy invested in creating it—has not yet been demonstrated. It is considered likely that scientific feasibility will be established for magnetic confinement, and perhaps for inertial confinement, within the next 5 years.

The panel recommends that our national program should continue to concentrate on the basic approaches without commitment to any single concept. Applications of fusion energy for purposes other than electricity generation should be analyzed in sufficient depth to obtain a meaningful assessment of the potential for alternatives uses. The move to pilot-plant experiments should not be attempted until a greater level of understanding is reached in the areas of confinement, plasma physics, and materials properties.

Fusion is far from the stage where it can be compared with other long-term energy systems, and its development to such a point will be costly. However, with continually improving scientific understanding and technological advances, achievable through a vigorously supported program, fusion may become increasingly attractive as an eventual long-range contributor to energy supply. It is important to continue international cooperation in this field, which has been singularly fruitful.

THE IMPACTS OF ALTERNATIVE APPROACHES ON PRODUCTION

In its scenario analysis, the panel examined each of the major energy sources to determine where the main supply problems lie. As suggested earlier, the supply problem is basically an oil and gas problem. For the reasons given in the previous discussions of these energy sources, the panel must recommend national commitments to synthetic substitutes. Otherwise, it appears that adequate supplies of other energy forms could be made available with enhanced supply scenarios. It is quite apparent that business as usual would not provide enough. Table 10 gives the panel's recommendations as to levels of commitment to individual energy sources.

The panel cautions against summing the contributions of the various energy sources. As noted previously, many of these estimates are based on undeveloped or partly developed technologies. The total for any future year, therefore, presumes that the energy technology for that source is fully developed, the assumed set of policies for each source are all in force, and the policies are effective. Although the panel has compiled these individual estimates as to what is possible for each source, it does not believe that all the sources will be developed equally successfully. Specifically, in the enhanced supply and national commitment scenarios, the technical confidence to be associated with the achievement of these targets varies considerably from source to source.

Government Incentives for Energy Supply

As stressed throughout this report, it is imperative that the federal government establish coordinated energy policies that will provide a basis for energy planning, for the lack of such policies discourages prospective investors. It is necessary to consider the respective roles of government and private industry in energy development.

Where two or more essentially developed technologies exist to meet a particular energy need (such as coal and LWRs for electricity production), the proper government role is to establish safety guidelines and environmental protection requirements, and then let a mix of supply technologies be determined by the market. Guidelines should be stable and specific so that potential suppliers of new capacity have reliable bases for planning. Such policies should be changed only when necessary to cover situations in which unforeseen discoveries (e.g., the problem of sulfate air pollution from coal) require it to protect the health and safety of the public. In a competitive atmosphere, there would be no need for subsidies, price controls, and tax incentives designed to encourage development of one technology over another or to restrict demand.

Table 10 Domestic energy production under recommended policies, in quads per year

Energy source	1975	1990	2000	2010	Recommended policy[a]
Crude oil (including natural gas liquids)	20.0	20.0	18.0	16.0	(E)
Syncrude from coal	--	0.7	4.7	12.9	(N)
Shale oil	--	2.0	2.5	3.0	(N)
Subtotal (liquid fuels)	(19.6)	(22.7)	(25.2)	(31.9)	
Natural gas	19.7	15.8	15.0	14.0	(E)
Synthetic gas from coal	--	1.7	4.5	7.9	(N)
Subtotal (gaseous fuels)	(19.4)	(17.5)	(19.5)	(21.9)	
Coal	16.4	26.6	37.2	49.5	(E)
Hydroelectric	2.4	4.1	5.0	5.0	(E)
Nuclear	2.7	13.0	29.5	41.7	(E)
Solar	--	1.7	5.9	10.7	(E)
Geothermal	--	0.6	1.6	4.1	(E)

[a] E = enhanced supply; N = national commitment.

Where substitute energy forms (such as synthetic crude oil from coal and shale) appear to be needed but are not now economically competitive, some incentive will be required to stimulate production. Government sharing of the costs of process development and plant demonstration is necessary because the investments are large, financial risks appreciable, and initial products not competitive. The government can make synthetic fuels more competitive by gradually deregulating the price of oil and natural gas. If these actions prove inadequate, it may be necessary to stimulate production of synthetic fuels by tax incentives or such measures as requiring a certain percentage (increasing with time) of delivered oil and gas to be of synthetic origin.

Advanced technologies such as solar energy in most forms, fission breeders, advanced forms of geothermal energy, and fusion are characterized by the need for an extended period of development before significant deployment could be achieved. In addition, the technical feasibility of fusion and the economic viability of breeders and advanced geothermal and solar concepts have yet to be proven. Part of the cost of research, development, and demonstration should be shared by government and industry. Tax incentives or government sharing of development costs are the most appropriate forms of government support. This panel does not favor taxing existing lower-cost energy sources to make new technologies competitive. Since several supply alternatives are likely to be available, there is little justification for consumers' bearing the cost of subsidizing any of these technologies on a long-term basis. Once the development effort has been completed, those technologies found to merit commercialization should compete with established energy supplies such as coal and nuclear without further government support.

The greatest percentage of government support for new technologies should be budgeted during the research and development stage, with industry gradually assuming the total burden as commercialization is approached. In order to develop the industrial capability for commercialization, the panel feels it is particularly important for the management of demonstration projects to be the primary responsibility of industry.

Energy Supply Contingencies

Much of the debate on the nontechnical aspects of energy has focused on the effect (both real and perceived) of energy growth on the environment, social structure, and domestic security of the United States. Of equal concern to the panel is the ability of the public to adjust to unexpected problems and emergencies, that could result from reliance on energy supplies that have insufficient margin and flexibility. A sudden or continuing disruption of energy supplies upon which much of the economy is dependent could have far-reaching effects, such as a reduction of industrial production, severe unemployment, business failures, adverse effects on health, and even civil unrest. A recent analysis by Decision Focus (1978) demonstrates that the total cost of a given shortfall of energy production capacity is substantially higher than that for an equal amount of excess capacity.

A gradual reduction in the ratio of energy consumption to gross national product is a desirable objective, provided it can be achieved through increased end-use efficiency (that is a reduction in Btus per dollar value), so as not to adversely affect the quality of life. Depletion of economically recoverable resources will eventually force such a reduction anyway, and such a transition should be effected as smoothly as possible.

Protection against energy interruptions can be provided through several means. Using a number of domestic energy forms rather than relying too heavily on one or two, for example, can lead to flexibility of supply and decreased reliance on imports. Interchangeability of fuels is also desirable. With regard to electricity production, maintaining a reserve capacity to accommodate abnormal peaking demand in addition to unexpectedly high equipment outages can minimize interruptions in service.

Storage of energy resources is also desirable to accommodate short-term interruptions of supply resulting from curtailments in imports, labor problems in energy production facilities, and disruptions to the domestic fuel delivery system caused by nature (such as those that occurred in the winters of 1976-77 and 1977-78), and the like.

In the longer term, after the turn of the century, the United States will have to rely on technologies that at this time are not well developed. It is important to recognize that not every one of these technologies is likely to be technologically and economically viable. It is thus imprudent to focus all development efforts on one or two alternatives. All promising technologies should be supported at least to the point at which their potential can be adequately assessed.

Lead Times in Deploying New Energy Systems

This report has commented on the desirability of diversifying our energy supply system and the need to facilitiate the development and deployment of new systems. In this regard, the panel recommends that attention be given to the time it takes to deploy new systems on a wide scale.

For established technologies, the total lead time consists of the time necessary to plan, design, site, and construct new facilities for resource development and energy production. For advanced sources it also includes the time necessary to develop and demonstrate the technology. Lead-time considerations also apply to the rate at which demand can be modified through the use of more energy-efficient facilities and equipment.

The starting point for implementation of any new technology is a decision by industry or government--or both--to take the steps that lead to large-scale commercial development. Generally the basic research and development would have been completed, although considerable supporting technology might still be under development, materials and equipment might be undergoing test, and so forth. The next step would be the design, construction, and operation of an experimental plant of a scale and character adequate to verify the technical features and to optimize the design parameters and establish the economics for a commercial plant (although economic operation may not be a requirement per se). More than one such plant may be required, and ordinarily these would be significantly close in design and scale to a commercial design. Commercial use of the technology could then expand at a rate determined by market forces until that energy form became a major contributor to total energy use.

The time it will take to complete this sequence of events for any given technology is difficult to predict. In practice, however, it tends to be underestimated, further emphasizing the need to establish energy policies that accommodate the long lead times required for new technologies and serving to remind us not to expect emerging energy technologies to play major roles in the short-term energy picture.

SUMMARY

The energy picture for the United States today is characterized by several factors that engendered heightened concern about management of all energy systems. All increase the risk to the U.S. economy and national security through possible interruptions in supplies of key energy sources. In addition, there is a persuasive argument that the trends of the past few years cannot continue and should, in fact, be reversed before they are changed by forces beyond the control of the United States.

While there are a variety of estimates of worldwide oil consumption, resources, reserves, and productive capacity, they generally result in a conclusion that even in the face of increasing demand and higher prices, the production from conventional oil sources will probably level out and peak during the period of 1990 to the year 2000, some 11 to 23 years from now.

The other industrialized countries—most of which do not even have the energy resources of the United States—will be seeking to increase their share of worldwide oil supplies. In addition, developing countries will be expanding more rapidly and adding to the pressure on diminishing supplies. Short of draconian measures which the United States would surely reject—it seems unlikely that the United States will be able to count on anything except steadily declining oil imports in the period from the early 1980s on.

Quite aside from these global considerations of supply and demand, the strain on our balance of payments, with an outflow of upwards of $50 billion per year close at hand, as well as the risks to our economy of importing over one quarter of all our energy needs (much of it increasingly from politically unstable areas) seem evident and necessitate strong and immediate action. The difficulty with that is that at this time there is nothing in terms of action underway in the United States to change these trends or to deal with the problems discussed in this introduction.

It must also be observed from the global point of view that the evident concern of the United States with respect to energy coupled with little visible action and even less accomplishments is destructive to U.S. world credibility and leadership. In particular, the failure of the United States to implement effective conservation measures, to make increasing and effective use of its nuclear resources and expertise, and its failure to utilize its disproportionately large coal resources are looked on by the rest of the world with virtual disbelief. Such a future does not bode well for international cooperation in the energy fields in the years to come.

REFERENCES

Decision Focus. 1978. Costs and Benefits of Over/Under Capacity in Electric Power System Planning. Prepared for the Electric Power Research Institute. Palo Alto, Calif.: Electric Power Research Institute (EA-927).

Energy Research and Development Administration. 1976a. Statistical Data of the Uranium Industry. Washington, D.C.: Energy Research and Development Administration (ERDA-6J0-0100).

Energy Research and Development Administration. 1976b. NURE Preliminary Report. Washington, D.C.: Energy Research and Development Administration (ERDA-6J0-111).

Energy Research and Development Administration. 1976c. Uranium Industry Seminar. Grand Junction Office. Washington, D.C.: Energy Research and Development Administration (ERDA-6J0-108), October 19, 20.

National Research Council. 1978. Problems of U.S. Uranium Resources and Supply to the Year 2010. Supporting Paper 1, Uranium Resource Group, Supply and Delivery Panel, Committee on Nuclear and Alternative Energy Systems. Washington, D.C.: National Academy of Sciences.

National Electric Reliability Council. 1977. 7th Annual Review of Overall Reliability and Adequacy of the North American Bulk Power Systems. Princeton, N.J.: National Electric Reliability Council.

U.S. Department of Energy. 1978. Monthly Energy Review. Washington, D.C.: U.S. Department of Energy (DOE/EIA-0035/6).

2 ELECTRICITY

In industrial societies energy is used in many forms for a variety of purposes. It most cases, the primary fuel must be converted to a secondary form for end use. Crude oil, for example, is refined into jet fuel or gasoline, and coal and uranium are used to generate electricity. Some fuels, such as coal and natural gas, can be consumed in essentially the forms in which they come from the ground, but almost every large-scale use of any fuel also depends directly or indirectly on electricity for controls, instrumentation, auxiliaries, or services. Electricity is used also, of course, to supply energy directly for many end uses such as heating, lighting, and power for motors.

The electric utility industry has grown rapidly since its beginning. Because it is a clean, convenient, and flexible energy form, electricity is highly desirable for industrial, residential, and commercial consuers. The American public's preference for electricity shows up in a growth rate in electricity consumption consistently higher than that of total energy consumption.

THE NATION'S ELECTRICAL SYSTEM

The electric power industry in the United States includes nearly 3,500 utility systems, which vary greatly in size, type of ownership, and range of functions. The industry is owned by a variety of investor-owned companies, nonfederal public agencies, cooperatives, and federal agencies, and is unique among worldwide systems in its diversity and complexity.

The 530 gigawatts of installed capacity that existed at the end of 1976 was about 76 percent investor-owned, about 20 percent under government or cooperative ownership, and about 4 percent belonging to industries that produce cogenerated power (Edison Electric, 1977). The capacity contribution by the government and cooperatives has remained at around 20 to 22 percent for the last 20 years, whereas the capacity of industry-produced cogenerated power has slipped from about 12 percent to 4 percent during this period. By the end of 1976, the total gross investment in the investor-owned segment of the electric utility industry alone was approximately $180 billion, representing investments larger than those in any other U.S. industry (Edison Electric, 1977).

In 1976, this country consumed 74.2 quadrillion Btu (quads) of primary energy equivalent. (In this estimate the Bureau of Mines convention of counting nuclear, hydroelectric, and geothermal energy in terms of the equivalent amount of coal to produce the same amount of electricity is used. Cf. footnote Table 1.) About 21 quads (or about 29 percent) were used to generate more than 2 trillion kilowatt-hours of electricity (U.S. Department of Energy, 1978b) distributed to over 82 million customers (Edison Electric, 1977). About 46 percent of this electricity was generated by coal, 18 percent by oil, 14 percent by gas; 12 percent by nuclear fuel; 10 percent by hydroelectric dams; and about 0.3 percent by geothermal energy, wood, and other small contributors (U.S. Department of Energy, 1978b). Figure 1 gives the electricity generation from 1920 through 1976 from all fuels, including electricity generation from nuclear power and from hydroelectric plants. The percentage of the nation's primary energy used to generate electricity grew from about 11 percent in 1920 to the current value of about 30 percent (U.S. Department of Energy, 1978b).

ELECTRICITY DEMAND GROWTH

Future growth in electricity demand will have an important bearing on the types of energy supply systems needed and on the appropriate combination of these systems needed to meet demand. Projections of future generating capacity are many and vary greatly, particularly over the longer term.

Near-Term Outlook

It takes 8 to 12 years to license and build a new base-load generating plant. Utilities therefore must make commitments for facilities well in advance of the actual need, as determined by load forecasts. The near-term generating capacity thus can be fairly firmly established. The Federal Energy Regulatory Commission (formerly the Federal Power Commission) and the National Electrical Reliability Council, among others, maintain compilations of industry plans and near-term projections for about 10 years into the future. As shown in Table 11, most projections for 1985 indicate an electric energy use of about 3.5 trillion kilowatt-hours, which represents about 37 quads of input energy and about 800 gigawatts of installed capacity (Edison Electric, 1976a; Federal Power, 1976; National Electric, 1977). This level of use

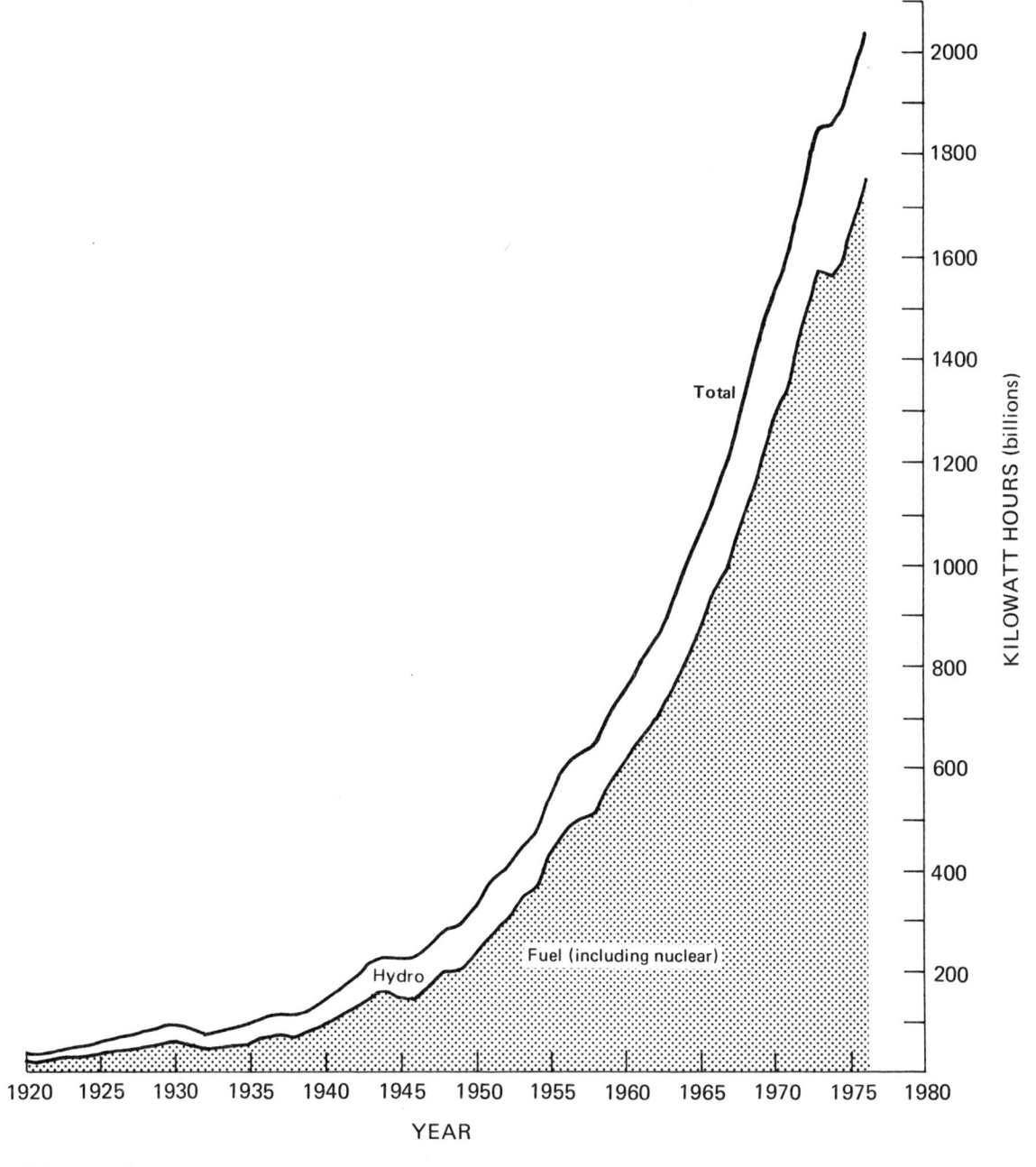

Figure 1 Electricity generation in the United States from 1920 to 1976 (Edison Electric, 1977).

reflects an average growth rate of 6.6 percent per year from the 1977 use of 2.1 trillion kilowatt-hours. The growth rate from 1956 to 1976 (Figure 2) shows wide variations in yearly totals with an average slightly more than 6 percent. Electricity use in 1977 was about 4.2 percent greater than in 1976.

Table 11 Projections of demand for input energy[a] for electricity generation, in quads per year

Study	1977	1985	1990	1995	2000	2010
Actual	22.5					
Edison Electric Institute (1976b)		34-39			51-86	
Electrical World (1977)		33	41	51		
Institute for Energy Analysis (1976)[b]		31-34			47-64	56-82
Ebasco Services (1977)		35	45	58		
U.S. Bureau of Mines (1975)		39			79	
National Electric Reliability Council (1977)		37				
Federal Power Commission (1976)		37				

[a] Input energy is the total energy consumed in producing electricity—not just the thermal value of the electricity produced. In general, the primary energy source, be it coal, oil, gas, or uranium, is consumed in the process of generating electricity, and only part of the energy content of the fuel is recovered as electrical energy. A pound of coal, if burned directly, might produce 9,000 Btu of heat energy, but if used to generate electricity it might produce only 1 kilowatt-hour, which is equivalent to 3,413 Btu. Thus, about 62 percent of the energy required to produce electricity is rejected waste heat. The ratio of the input primary energy to the output secondary electrical energy is a measure of the thermodynamic efficiency of the electric generating system. This efficiency for present-day power stations varies from as low as 20 or 25 percent for gas-turbine peaking plants to around 38 or 39 percent for the best modern coal-fired plants. The national average for fossil-fueled steam electric plants is around 32 to 34 percent. In this table, an efficiency of 32.5 percent was used to convert estimates given in kilowatt-hours to quads.

[b] Includes more than one growth case.

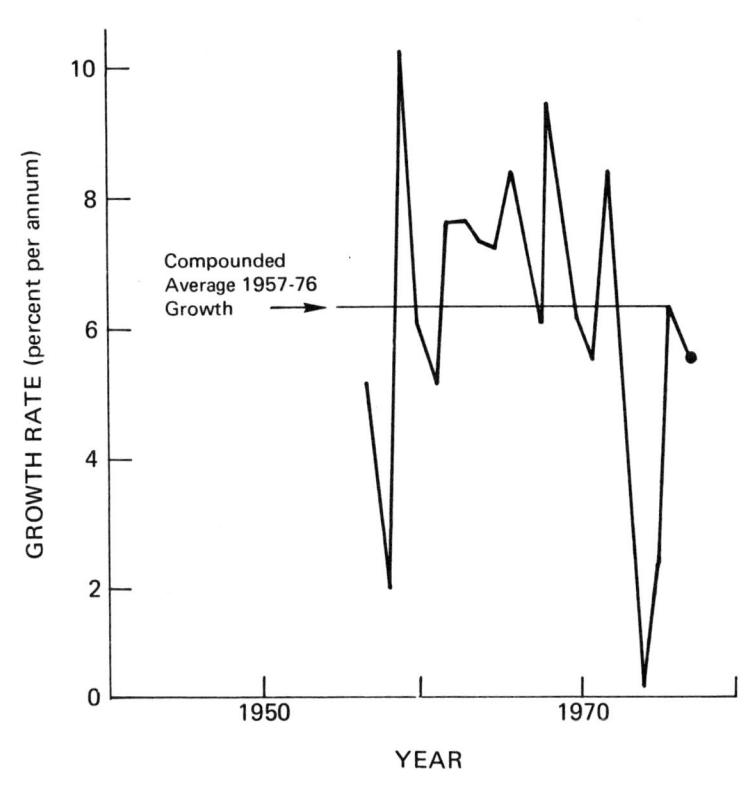

Figure 2 Annual growth rates of electricity generation from 1957 to 1977 (Edison Electric, 1967a; Electrical World, 1977; U.S. Department of Energy, 1979).

Long-Term Outlook

In the longer term, the referenced studies vary much more. An Institute for Energy Analysis (1976) study gave 2010 estimates, but most have projections for no further in the future than 1995 or 2000. In these studies, the growth rates are positive, although they decrease with time. Several of these studies also projected total energy demand; it is interesting to note that energy demand for electricity rises as a percentage of total energy demand in each projection (Table 12), approaching nearly 50 percent after the year 2000. In those studies with more than one case (Edison Electric Institute and Institute for Energy Analysis), the percentage of electricity is projected to be about the same whether total energy demand is high or low (in 2000, from 100 to 180 quads per year).

Table 12 Energy demand for electricity generation, as percentage of total energy demand

Source	1977	1985	1990	1995	2000	2010
Edison Electric Institute (1976b)	30				40-45	
Institute for Energy Analysis (1976)	30	38-39			47-51	47-52
Ebasco Services (1977)	30	37	42	49		
U.S. Bureau of Mines (1975)	30	38			48	

The many advantages of electricity for the consumer--and the motivation to substitute it where practical for uncertain oil and gas supplies--are strong reasons for believing that it will continue to grow faster than total energy use for some time into the future.

ELECTRICITY SUPPLY SCENARIOS

In the CONAES study, demand projections were conducted by the Demand and Conservation Panel (National Research Council, 1979) and the Synthesis Panel's Modeling Resource Group (National Research Council, 1978) estimated the elasticity of electricity demand with respect to energy prices. These reports should be consulted for descriptions of methodology and results. The Supply and Delivery Panel did not specifically assess the demand for electricity. Its task was rather to estimate the availability of primary energy resources under three scenarios, or sets of institutional and political conditions--described as business-as-usual, enhanced supply, and national commitment--for the years leading up to 2010. (Chapter 1 describes this scenario analysis.) The panel then proportioned the scenario estimates of primary resources into the forms of secondary energy expected to be used (liquids, gases, electricity, and heat), to obtain estimates of how much of the various resources would be available for electricity generation in the three scenarios (see Figure 3).

The Near Term

For 1985, the amounts of input energy estimated to be available domestically for electricity generation in the business-as-usual scenario is 26 quads; for the enhanced supply scenario it is 32 quads; and for the national commitment scenario it is 37 quads. Thus, even in the short term, a business-as-usual approach to developing energy supply is likely to produce less input energy than will be needed to meet the demand projections summarized in Table 11, and the enhanced supply approach gives estimates near the low end of the range. The national commitment scenario total of 37 quads lies comfortably in the range of projections, but because it would require an immediate national commitment to producing almost all domestic energy resources it is not a realistic goal. This means that a few quads of imported energy will continue to be necessary in electricity generation to meet these rather firm demand projections.

The Long Term

As noted earlier, electricity demand projections for the years after 1990 show wide variance. The estimates of electrical demand in Table 11 for the year 2000 show energy requirements in excess of 50 quads per year. This is much higher than this panel's estimate of supply in 2010 under business-as-usual conditions (36 quads per year), and somewhat higher than the estimates for enhanced supply conditions (47 quads per year).

In summary, most of the studies cited seem to center around a yearly growth in electricity demand of about 4 percent per year for the period through the year 2010. Although the annual rate was between 6 and 7 percent before 1973 and declined to near zero in 1974, it was 4.2 percent in 1977; and for the first nine months in 1978 it was about 3.3 percent. From this past trajectory, most projections indicate a lowered (but positive) growth rate past the turn of the century to a value near 3 percent by 2010, requiring more than a tripling of electrical generating capacity by 2010 (Figure 3).

Although these projections could be in error in either direction, the panel believes that an annual rate of about 6 percent is likely to be an upper limit and a doubling of capacity by 2010 (2 percent per year) a lower limit, with either considered unlikely. The panel estimates of

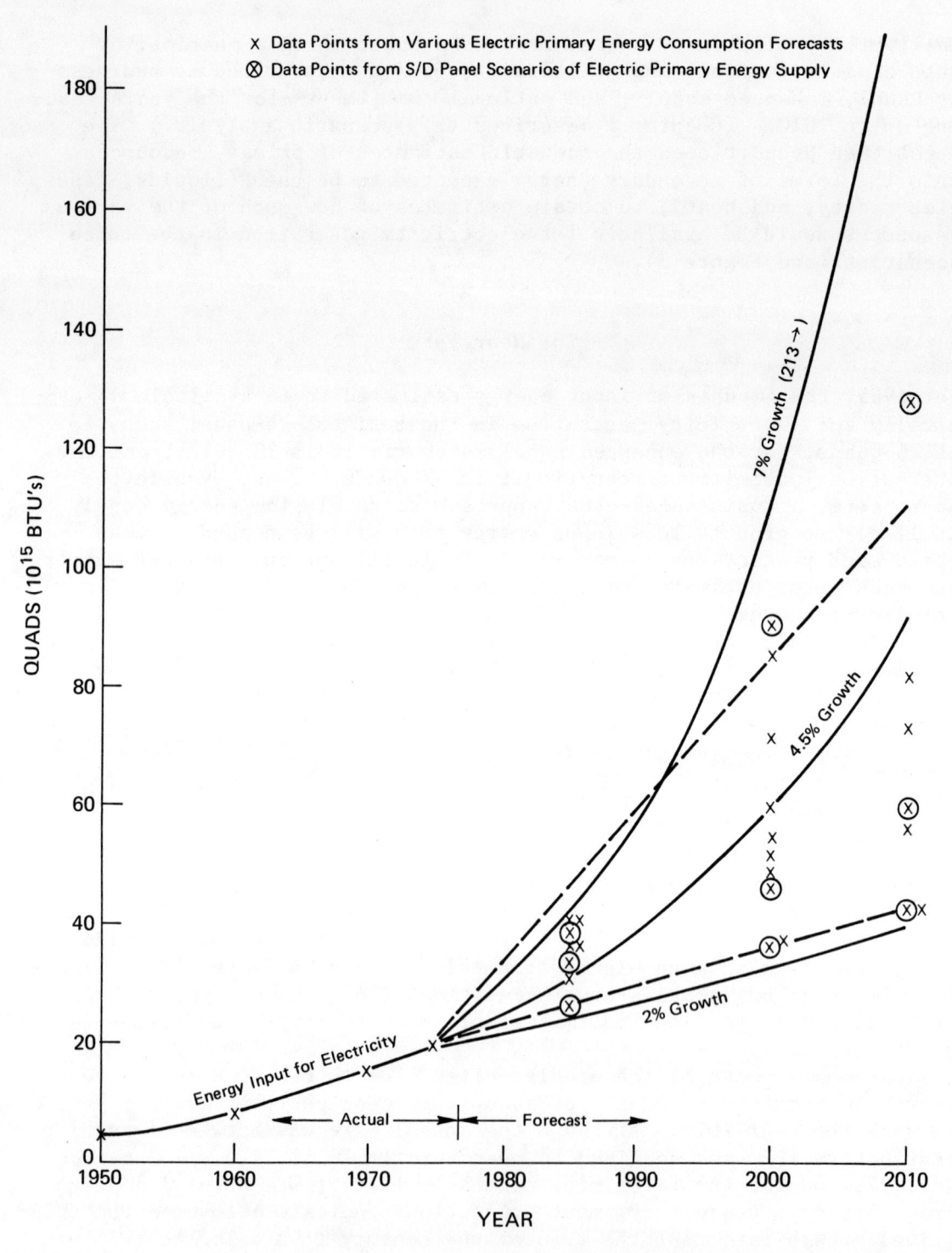

Figure 3 Envelope of forecasts of energy requirements for electricity generation, in quads per year. Points marked "x" from Edison Electric (1976b) and Electrical World (1977); points marked with circled "x" from Supply and Delivery Panel scenarios of primary energy supply for electricity.

input energy supply in Table 11 show that under the business-as-usual scenario, there would be barely enough electricity generated at the lower limits and that to meet projected demand an input energy supply based on enhanced supply or national commitment supply policies would be needed.

Fuel Mix to Meet Electricity Needs

The changes in the fuel mix for electricity generation from 1961 to 1975 (Figure 4) clearly demonstrate the emergence of nuclear power since 1970. It also shows the increased use of oil and gas in the late 1960s in response to the need for stricter environmental control. The expected contributions of individual energy sources will be discussed later, but it is generally projected that future base-load generation will need to rely almost completely on coal and nuclear power. After 1985, the use of oil will diminish as present plants are retired. Gas is already playing a steadily declining role; it will probably be essentially phased out by the year 2000. Hydroelectric generation will increase only slightly if at all, mainly because of the lack of suitable sites and the environmental impacts of building new dams, but there is at least one estimate that the installation of additional generating capacity at existing dam sites could add as much as 54 GWe to the nation's power pool by the year 2000 (Lilienthal, 1977). There is little doubt of what the fuel mix will be for the near term, because it is based on utility plans already under way. Figure 5, taken from a National Electrical Reliability Council report, forecasts electricity generation by principal energy source for 1977 and 1986.

The additional electrical generation necessary for meeting peak loads —roughly 10 percent of capacity and from 2 to 5 percent of the system energy production—will be derived from small (several-hundred-megawatt) coal-fired plants, pumped hydroelectric or compressed air storage, combustion turbines that will ultimately use medium-Btu gas generated from coal on site, synthetic liquid fuels; and possibly closed-cycle gas turbines driven by direct combustion of coal with high-temperature heat exchangers. As new base-load capacity is added, existing oil- and gas-fired units and those under construction will be kept on line to meet peaking and intermediate requirements. Existing oil- and gas-fired units constitute almost 150 GWe of capacity (Electricity Conversion, 1977) and have a replacement value of about $60 billion.

An Edison Electric Institute (1976a) report supports the Supply and Delivery Panel's conclusion that the prime energy sources for electricity in the next few decades will be coal and nuclear power. Uncertainties, however, were too great to allow sharp conclusions in that report on the split between coal and nuclear energy. Figure 6 is an attempt to project the composite plans of the utility industry (through 1985) and the Edison Electric Institute projections for the year 2000.

The Institute for Energy Analysis (1976) made high-coal and high-nuclear projections corresponding to both low-demand and high-demand energy consumption estimates (Figure 7). Their results cannot be compared directly with the Edision Electric Institute projections, since

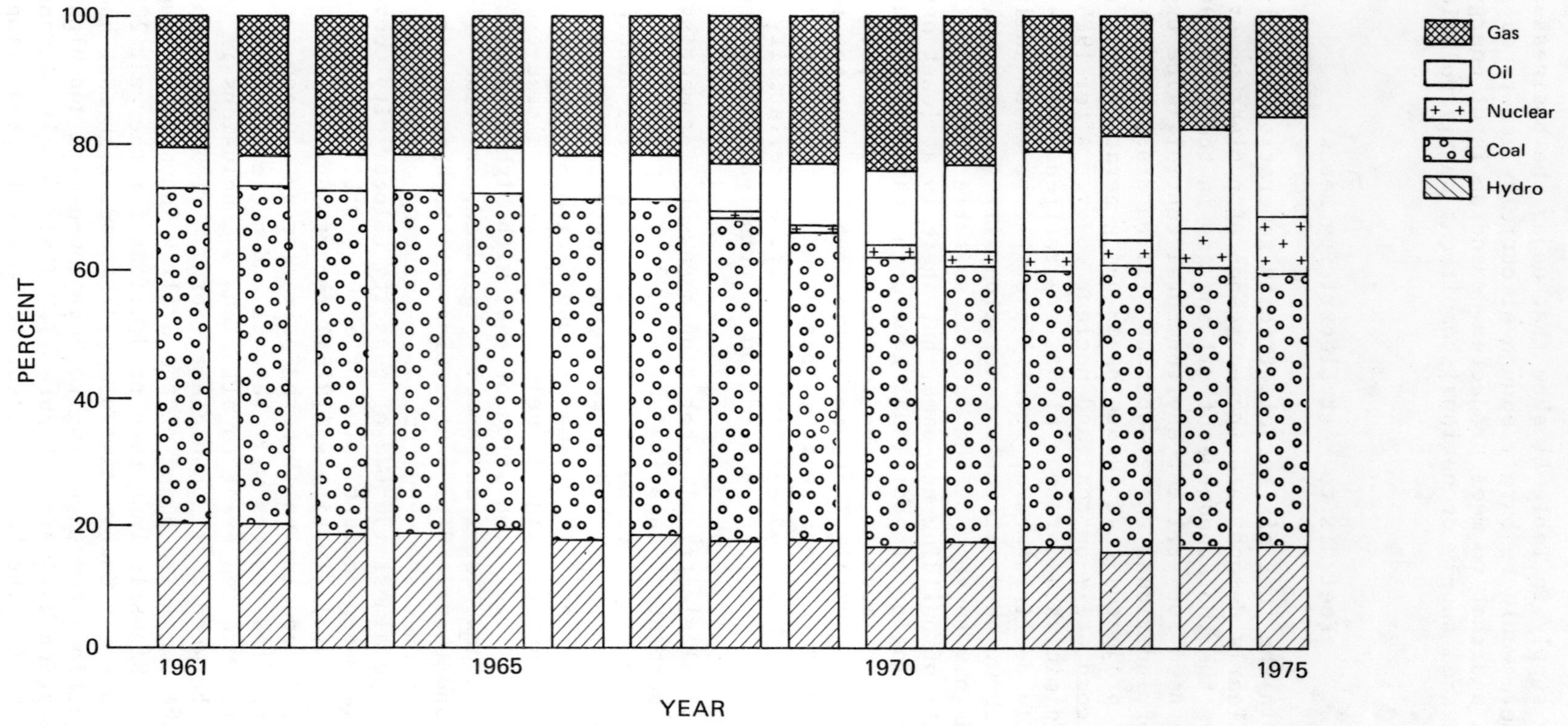

Figure 4 Energy sources for electricity generation from 1961 to 1975 (Electricity Conversion, 1977).

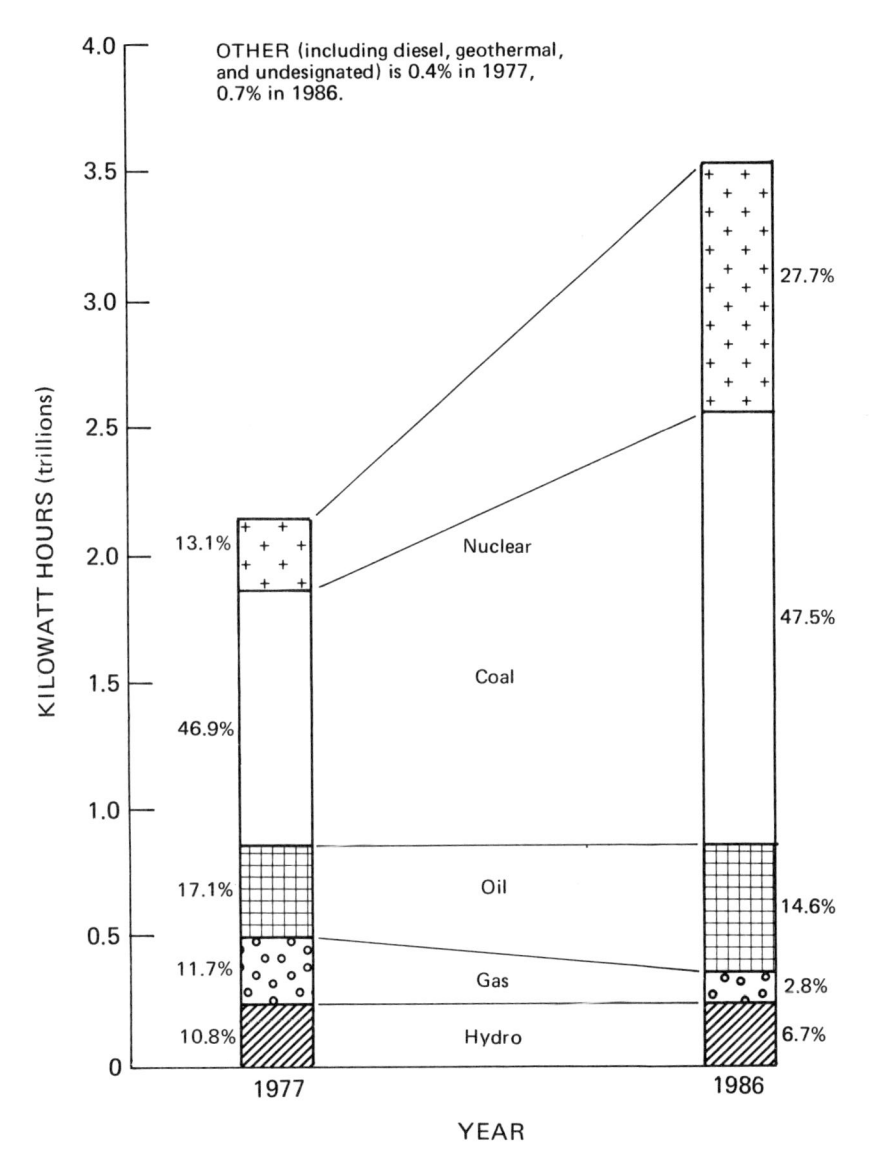

Figure 5 Forecast of electric generation from 1977 to 1986 for the contiguous United States, by principal energy source (National Electric, 1977).

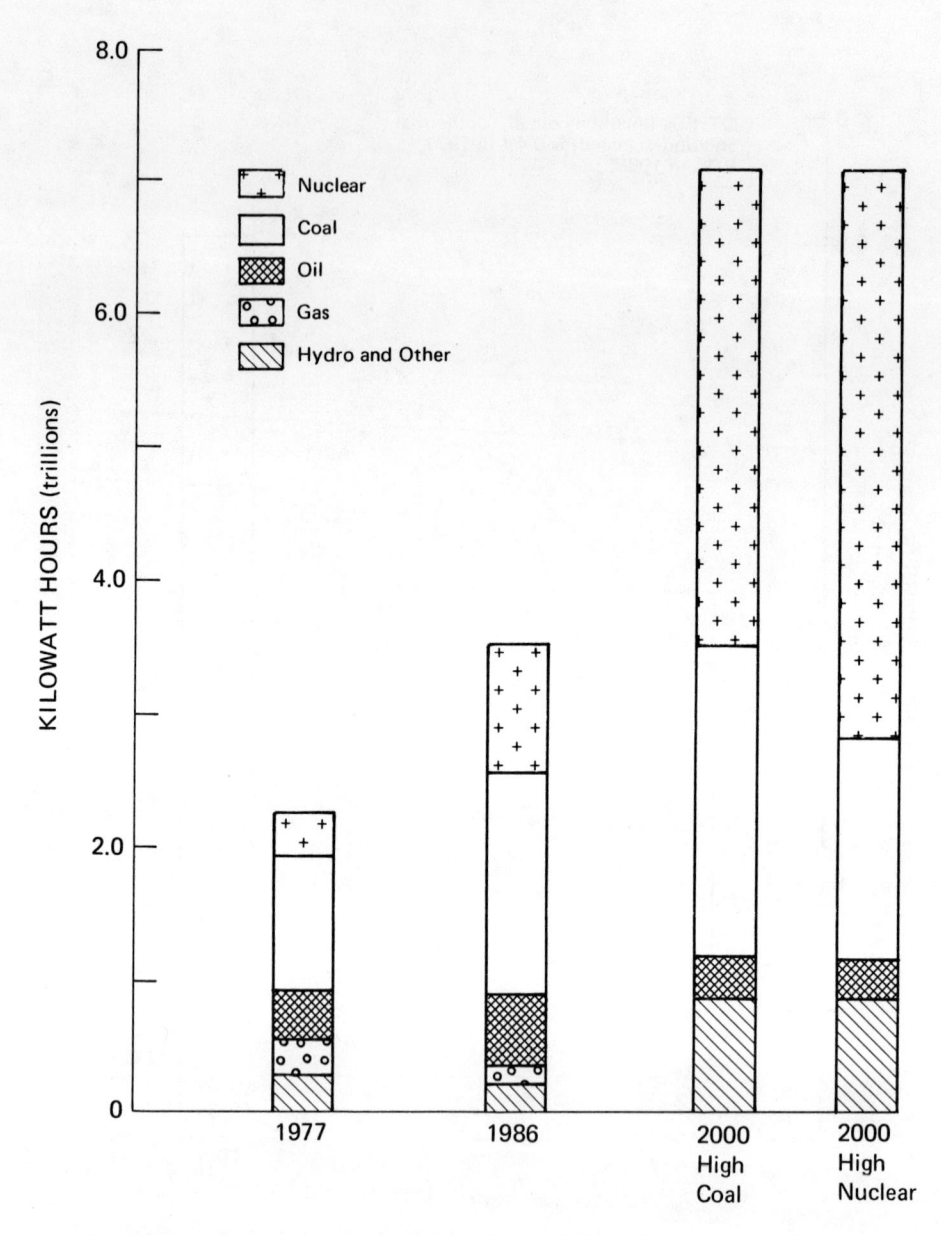

Figure 6 Electricity generation for 1977, 1986 and 2000, by energy
source, showing possible tradeoff between coal and nuclear
fission in 2000 (National Electric, 1977 for 1977 and 1986;
Edison Electric, 1976b for 2000).

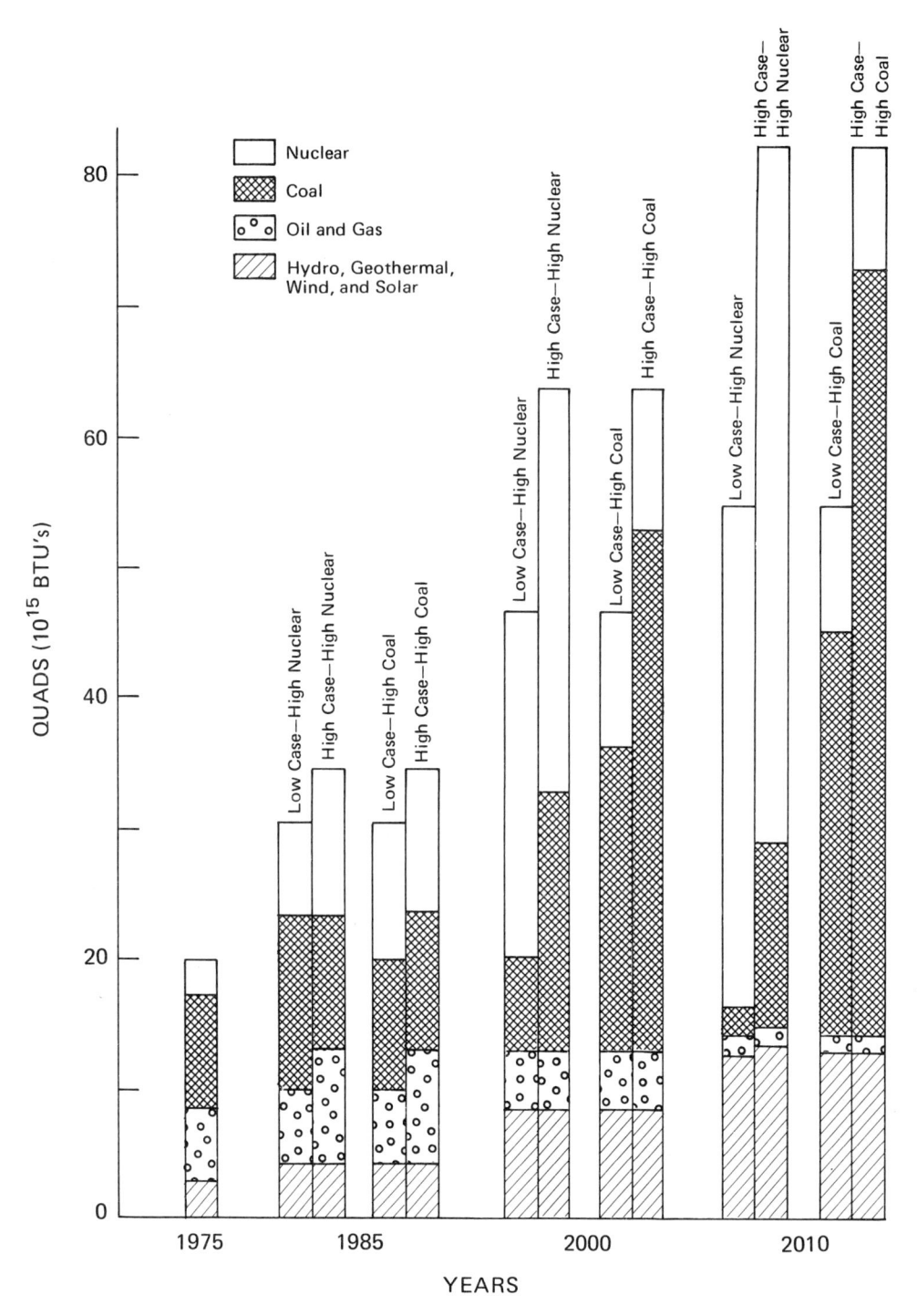

Figure 7 Summary of energy inputs to electricity for 1975, 1985, 2000 and 2010, by source, showing coal and nuclear tradeoffs for both high- and low-demand projections (Institute for Energy Analysis, 1976).

the high-coal (low-nuclear) case was based on the assumption of a nuclear moratorium, with nuclear energy inputs to electricity remaining unchanged from the 1985 level of 10.6 quads. However, the low-demand, high-nuclear, and high-demand, high-nuclear fuel mix cases are probably more plausible. In these examples nuclear energy inputs for the year 2000 vary from 27 to 31 quads.

The costs of electricity from coal-fired and nuclear plants are roughly comparable, with regional variations and regulatory uncertainties. In regions of the country with ready access to coal, coal-fired plants tend to be more economical. In areas removed from easy access to coal, however, nuclear power tends to be more economical. (See Section entitled "Nuclear/Coal Cost Comparison," later in this chapter.)

The capital costs of nuclear plants are about 15 to 25 percent greater than those of coal-fired units, per kilowatt of capacity. However, lower fuel costs make nuclear power less sensitive to future inflation. The capital costs of a new coal-fired plant are heavily (more than 20 percent) burdened by the costs of meeting current environmental requirements and are vulnerable to further cost increases, beyond general inflation, which may result from imposition of more stringent environmental requirements.

The operating costs of fossil-fired power plants (largely fuel costs) are a significantly larger part of total generating costs than those of nuclear power plants, and are also more vulnerable to inflation over the plant's lifetime. For example, recent fossil fuel price increases have had large impacts on power costs, so that projected future fuel price trends will strongly influence utilities' choices among the available types of plants. Although uranium prices have undergone similar dramatic increases, these have had a smaller effect on power costs because the share of nuclear fuel in the total cost of producing power is much smaller than the share of fossil fuel. Also, uranium price increases appear to be moderating in response to reduced nuclear growth projections and expanded exploration.

Although capacity and availability factors of coal and nuclear baseload plants are comparable (Atomic Industrial, 1977), it is clear that choices between the two should be governed by more than economics. A balanced dependence on coal and nuclear energy will provide valuable diversity, and therefore resilience in the face of single-fuel supply interruptions. Also, it does not appear likely that either coal or nuclear power stations alone could meet the total demand for additional capacity.

Figure 8 portrays a series of reasonable estimates of fuel sources available to meet an average electricity consumption of 7 trillion kilowatt-hours per year in the period from 2000 to 2010. Seven trillion kilowatt-hours, which amounts to 70 quads of energy input, represents a moderate forecast of growth in electricity demand (about 4 percent per year). (See Table 11 and Table 12.) If an estimated 1 trillion kilowatt-hours are to be supplied by hydroelectric, oil, gas, wind, tide, solar, geothermal, and so forth, the remaining 6 trillion kilowatt-hours must be supplied by coal or nuclear power. Figure 8 shows that under the business-as-usual scenario if certain limits are imposed on nuclear generation, such as no fuel recycling, nuclear power

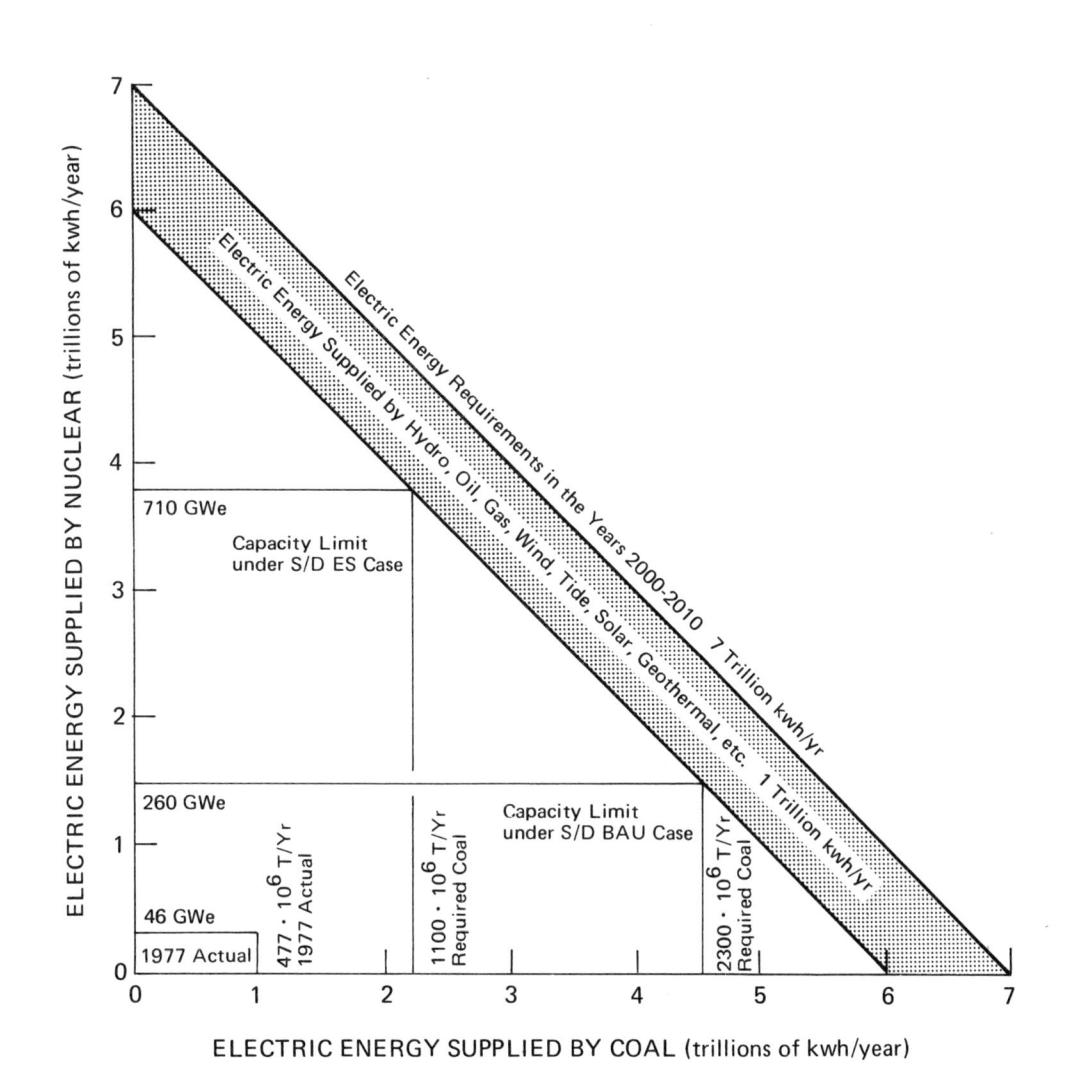

Figure 8   Estimates of fuel sources available for electricity generation in the period 2000 to 2010, illustrating a range of possible combinations of coal-fired and nuclear capacity.

can provide approximately 1.4 trillion kilowatt-hours per year. The remaining 4.6 trillion kilowatt-hours, to be provided by coal, will require 2.3 billion tons of coal each year just to meet electricity requirements.

In 1977 electricity generation consumed 477 million tons of coal (about 65 percent of total coal production). Thus, meeting such a large need for generating capacity solely with coal-fired plants would require the almost impossible feat of increasing net mine output more than fourfold by 2010. The replacement of depleted mines over the period, would amount to an additional 15 million tons per year, leading to a fivefold or sixfold increase in the rate of mine development. It is clear that if these projections of electricity demand are correct, current restrictions on nuclear power must be relieved to permit a more realistic balance between nuclear and coal-fired capacity.

The foregoing discussion implies that the national electricity system should plan for (a) declining use of gas and oil, increasingly restricted to peaking; (b) slowly expanded use of hydroelectric power (including pumped storage); (c) stronger dependence on coal and nuclear energy to meet base-load needs and some intermediate-load needs; (d) little near-term contribution from advanced fuel sources; and (e) prompt resumption of the breeder reactor program to ensure continued growth in the use of nuclear energy for electricity generation as highgrade uranium resources are depleted (see Chapter 5).

COAL FOR ELECTRICITY GENERATION

Past and Future Uses

Coal has been and, for at least the next decade or so, will continue to be the largest source of fuel for electricity generating stations in the United States. Coal was used to generate 57 percent of the country's electricity in 1925 (National Coal, 1972) and 46 percent in 1976 (U.S. Department of Energy, 1978b). It averaged 52 percent from 1920 to 1970 (National Coal, 1972). Although coal's percentage share of the electricity generation market has declined gradually during this period, the amount of coal used for this purpose has increased. About 177 million tons of coal was burned in 1960 to produce 400 billion kilowatt-hours of electricity (54 percent of the total) (Edison Electric, 1977; National Coal, 1972). In 1976, 448 million tons of coal were burned to produce 945 billion kilowatt-hours (U.S. Department of Energy, 1978b). Thus, during a period when coal's share of the market declined 8 percent, the amount consumed for generating electricity more than doubled. In recent years, electric utilities have burned about 65 percent of all the coal mined in the United States and they will continue to be the nation's major coal consumers.

According to utility plans for adding generating capacity up to 1985, coal will continue to fuel about 47 percent of electricity generation. This will require annual coal consumption by 1985 of more than 875 million tons merely to supply the electric power industry, with a total production of 1.3 billion tons to meet the added needs of other coal users. Figure 9 shows these requirements.

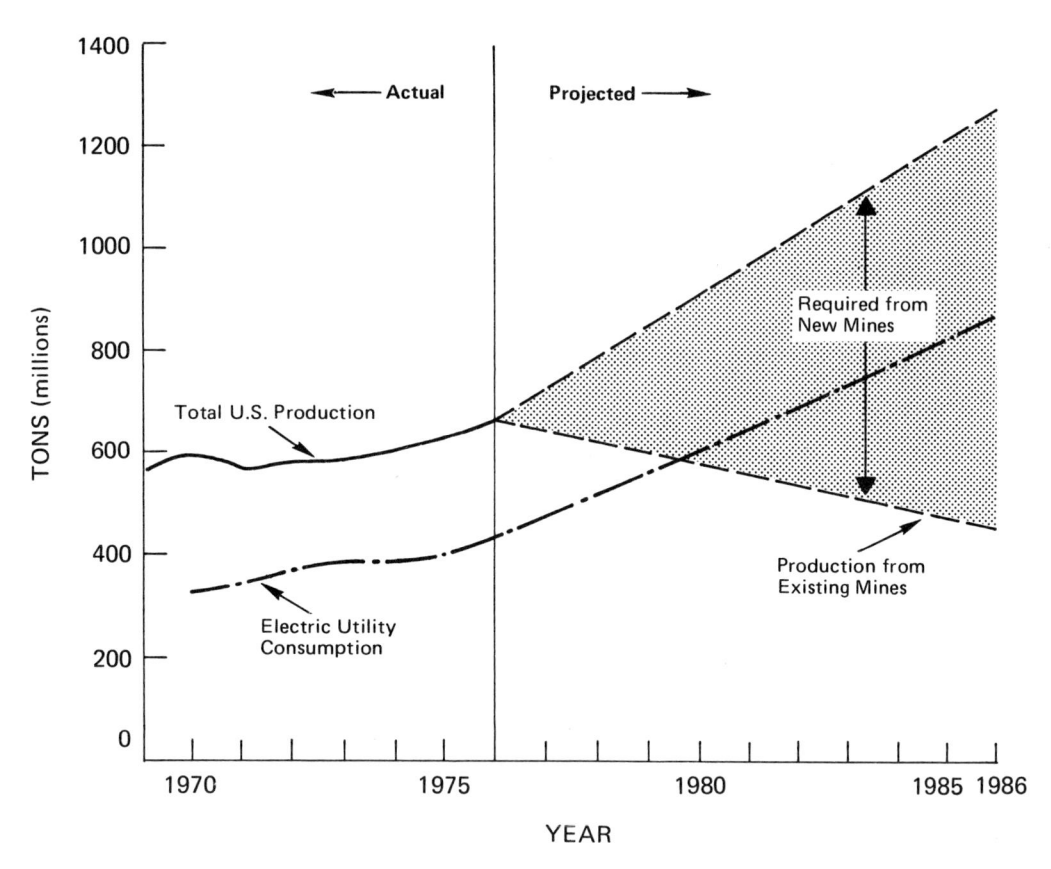

Figure 9 Actual and projected coal production requirements from 1970 to 1986 based on experience (1970-1976) and utility forecasts (1977-1986) (National Electric, 1977).

In September 1977, the journal Electrical World (1977) concluded that utilities had delayed their orders of nuclear plants for so long that, because of the required nuclear lead time, coal and to some extent oil were the only options available that could provide energy for capacity due on line in 1986 and 1987. In the meantime, however, very few orders for coal or nuclear plants have been placed, and demand for coal is not expected to strain the available supplies in this period.

The longer term needs for coal for electricity generation are more difficult to estimate. The high-coal and high-nuclear cases shown in Figure 6 for the year 2000 would require respectively 1.2 billion and 850 million tons of coal annually. These figures correspond approximately to the Supply and Delivery Panel's business-as-usual and enhanced supply scenarios for coal within the uncertainties of the estimates. The panel's figures increase only slightly for the year 2010 on the assumption that coal will be used increasingly to replace oil and gas and that nuclear power and the advanced solar and geothermal technologies will assume greater shares of the electricity generating load.

Geographical Distribution of Coal

The locations and qualities of coal deposits and the applicable mining methods are vital to determining how competitive coal can continue to be compared with other fuels used for electricity generation. Coal deposits underlie about 459,000 square miles in 37 states (Averitt, 1975), and major reserves exist near many large centers of industry and population. On a tonnage basis, these reserves are located almost equally east and west of the Mississippi River. On an energy basis, though, over 55 percent is east of the Mississippi.

Although 60 percent of the nation's coal reserves have sulfur contents of 1 percent by weight, only a small portion of the reserves were able to meet the strict standards established by the U.S. Environmental Protection Agency in the Clean Air Act of 1970 (Federal Energy Administration, 1974). The Clean Air Act Amendments of 1977 require scrubbers no matter what the initial sulfur content of the coal. Much of the coal reserves are located in the mountains of the West, in remote areas that are sparsely populated. Thus, to be used for the generation of electricity, these reserves must be transported long distances in the form of coal, synthetic liquid fuels, or generated electricity. There are--for the use of coal in electricity generation--added economic penalties for either transportation or constructing and operating desulfurization equipment, which must be considered in comparing the cost of coal with the cost of alternative systems such as nuclear power.

Another factor to be considered is the method of production--for example, underground versus surface mining. Surface mining is generally cheaper where it can be applied. Most of the cheap surface-minable coal reserves are in the western United States, long distances from electricity markets.

Pollution Control Technology

Air quality standards are the main environmental restraint on expanding the use of coal. In the future, increased use of coal may prompt increasingly stringent environmental regulations that will probably require major investments in control facilities. Sulfur dioxide emissions are the pollutant upon which air quality standards now concentrate. A number of other pollutants are produced in burning coal and this panel did not deal specifically with them. The following discussion of sulfur emission control is intended merely as a summary of the economic and technical problems of pollution control. (See the Risk and Impact Panel report for a fuller treatment of these and other environmental issues.)

The best available method for removing sulfur dioxide from power plant flue gases is a wet lime or limestone scrubbing process. It is relatively inefficient and expensive and carries its own pollution problems. It adds about 20 percent to the capital cost of the plant, roughly doubles operating and maintenance costs (excluding fuel costs), and reduces plant capacity by about 5 percent because of its energy consumption. It also lowers the reliability of the plant, so that overall plant availability falls. Finally, it produces large amounts of wet calcium sulfate sludge, in quantity that is about equal to the ash that would normally be produced; but it presents a much more difficult environmental isolation problem. There are strong incentives to develop more economical and environmentally benign sulfur removal processes, but much work remains to be done before these processes can be commercially practical.

Synthetic liquid and gaseous fuels from coal offer a better prospect for clean fuels. The energy lost in conversion and the high capital cost of the conversion facilities, however, increase fuel costs, and consequently, power generation costs to levels unlikely to be competitive with direct use of coal. Use of a combined cycle (that is, a combustion turbine with waste heat recovery in a steam cycle) can regain lost energy efficiency, but at the expense of higher capital costs.

The fluidized-bed combustion process, in which the oxides of sulfur react with limestone in the combustion stage rather than in the flue gas, offers promise as an economically competitive approach to coal use. Pilot demonstration programs are underway, but significant use will probably not be possible before the late 1990s. No coal burning technology on the horizon offers a major breakthrough in burning coal cleanly.

OIL AND GAS FOR ELECTRICITY GENERATION

Throughout the preceding discussions, it has been stressed that current concerns are greatly influenced by declining supplies of domestic oil and gas. The United States' stated policy is to move from the use of oil and gas for electricity generation.

Oil-Fired Electricity Generation

The annual amount of electricity generated by burning oil increased (with only two interruptions) from 28 billion kilowatt-hours in 1951 to 313 billion kilowatt-hours in 1973 (Edison Electric, 1977). Generation by oil dropped slightly in 1974 and 1975 to about 299 and 290 billion kilowatt-hours respectively, but in 1976 it jumped to an all-time high of about 320 billion kilowatt-hours, requiring 556 million barrels of oil (Edison Electric, 1977; U.S. Department of Energy, 1978b). The National Electric Reliability Council (1977) (NERC) indicates that completion of new oil-fired generating units already committed to construction will raise the industry's oil requirements to nearly 900 million barrels per year by 1982, with little or no further increase through 1986. By that time, according to the Federal Power Commission (1976), oil will be needed for only about 13 percent (NERC estimates nearly 15 percent) of the total electricity generated (465 kilowatt-hours), as compared with nearly 17 percent (313 billion kilowatt-hours) in 1973 and again in 1977.

The future supply of oil for the United States, including that for electric utilities, is highly dependent on imports. This is especially true in the Northeast, which has the highest fraction of oil-fired generating capacity and the greatest dependence on oil imports. Thus, the vulnerability of oil supplies and the increasing and unpredictable cost of oil are two factors influencing the electric utility industry to move as rapidly as possible to coal-fired and nuclear plants for future base-load generation. Also, the current United States policy prohibits construction of new oil-fired capacity.

As coal and nuclear plants assume a greater share of base-load generation, oil plants will be increasingly used for peaking or load-following purposes. Thus, beyond about 1985, the use of oil for electricity generation is expected to decline, reaching less than half its 1985 level by the year 2000 and continuing to decline thereafter. However, utilities will probably continue using substantial amounts of oil well into the next century.

Gas-Fired Electricity Generation

The use of gas for electricity generation grew steadily from 57 billion kilowatt-hours in 1951 to a peak of 376 billion kilowatt-hours in 1971 (Edison Electric, 1977), with its share of the market during this period ranging from about 21 percent in 1951 to a peak of about 29 percent in 1970. The use of gas for electricity generation declined rapidly after about 1971, generating only about 293 billion kilowatt-hours (14 percent of the nation's electricity) in 1976 at the expense of 3 trillion cubic feet of gas (Edison Electric, 1977; U.S. Department of Energy, 1978a).

As to the future of gas for the generation of electricity, the NERC study (see Figure 5) estimates that by 1986 gas will be used to generate only about 110 billion kilowatt-hours, requiring slightly more than 1 trillion cubic feet of gas (an input of about 1 quad) (National

Electric 1977). The Federal Power Commission (1976) estimate for 1985 is about double the NERC estimate. This may seem a wide variation, but there is great uncertainty in estimates of the future availability of natural gas. Utilities, however, have moved since 1972 to reduce their dependence on this fuel, and it appears that its use in generating electricity will be negligible by the year 2000.

NUCLEAR ENERGY FOR ELECTRICITY GENERATION

Present Status

At present, the predominant civilian role for nuclear energy is electricity generation. It has potential as a source of industrial process heat, but this has yet to be realized on a commercial scale and will probably be significant only if high-temperature gas-cooled reactors are commercialized and achieve a high market penetration. This appears unlikely under present conditions. Consequently, nuclear energy's role depends mainly on the future need for electricity and on the availability and competitiveness of other fuels.

In 1951, the Experimental Fast Breeder Reactor No. 1 generated the world's first electric power from nuclear energy. By early 1978, there were 69 commercial nuclear reactors with operating licenses in the United States, with a total capacity of 50 gigawatts (nearly 9 percent of U.S. generating capacity) (U.S. Department of Energy, 1979). During 1977, these reactors generated 251 billion kilowatt-hours. Growth in nuclear generation will continue until the lack of orders experienced over the past few years manifests itself in a shortage of plants coming on line in the late 1980s.

Future Role of Nuclear Power

As of February 28, 1978, there were 149 nuclear reactors on order in the United States with a total capacity of about 163 gigawatts (Kidder, Peabody, 1977). Most of these reactors are scheduled to be in operation by 1990.

Since, given existing regulations, it takes 10 to 12 years or longer to obtain site approval and to license, design, construct, and bring into operation a new reactor, the installed capacity for nuclear power plants is already established for the period up to about 1990 (Lester, 1978). Short of a concentrated national effort to accelerate the licensing of nuclear plants, this is not likely to change. This section will be concerned with the period from 1990 to 2010 and beyond.

Figure 10 gives the nuclear electric generating capacity operating and ordered in the United States through 1977. Although initial growth was rapid, orders have declined precipitously since 1974 because of the regulatory, financial, and political uncertainties confronting nuclear power, and there is not evidence yet of a resumption of the former order rate.

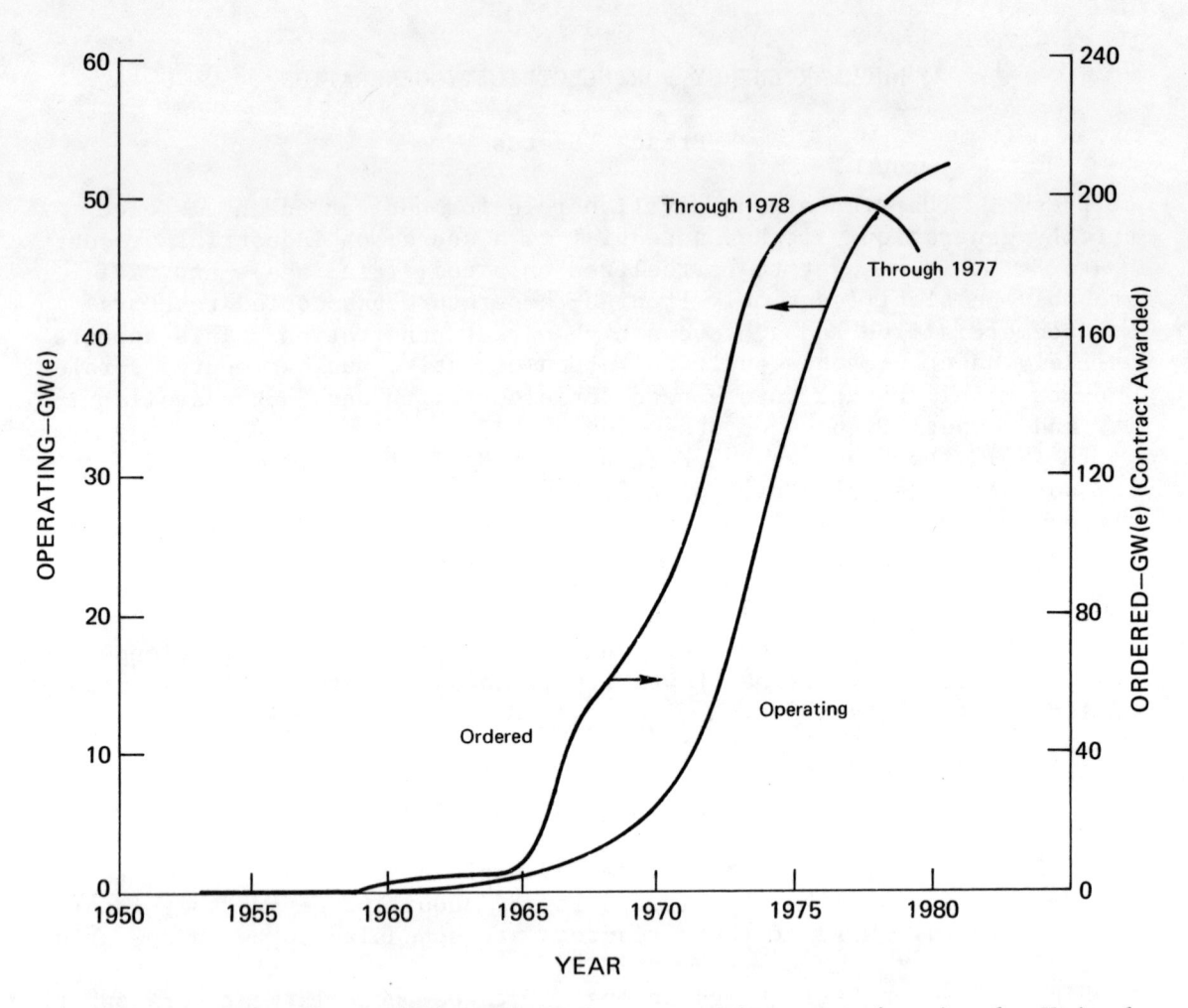

Figure 10 Operating nuclear electric generating capacity in the United States through 1977, and capacity on order through 1978 (Kidder, Peabody, and Co., 1978).

The Supply and Delivery Panel believes that the United States should plan to have nuclear power plants supplying about half of its base-load generating capacity by the year 2000. Indications are that fuels other than coal and uranium will supply about 20 percent of the requirements; coal and nuclear power should divide the remaining 80 percent equally. If, as described earlier, a load growth of 4 percent per year is reached, electricity production would require 60 quads of input energy and 1,500 gigawatts of generating capacity in the year 2000. This would require 600 gigawatts of nuclear plants in the year 2000 and more than 900 gigawatts by the year 2010. Unfortunately, barring a national commitment to nuclear power it is already too late to meet this schedule. If there were what the panel believes would be an unreasonably low growth of 2 percent per year, 325 gigawatts of nuclear capacity would be required in the year 2000 and 400 gigawatts by 2010, according to the proportions outlined earlier.

A more realistic goal, but still a very difficult one, would be to strive for a nuclear capacity of about 500 gigawatts by the year 2000 and 750 gigawatts by 2010. On the basis of existing orders, we are likely to have about 200 gigawatts on line by 1990. Achieving the desired levels of nuclear capacity, therefore, calls for a continuation of the growth that prevailed until 1974, namely, placing an average of about 25 gigawatts in new orders per year. This schedule is within the capacity of the industry, but realizing it depends heavily on (1) a more favorable financial, political, and regulatory climate than has prevailed for the past few years, and (2) an early reversal of the current stagnation in the nuclear business.

HYDROELECTRIC POWER

Hydroelectric power capacity grew from about 5 gigawatts in 1920 (Federal Power, 1971, 1976) to 59 gigawatts by 1976 (National Electric, 1977). Most early hydroelectric plants were used to satisfy base-load electrical needs. Thirty-five years ago it constituted 30 percent of U.S. generating capacity and supplied 40 percent of the electricity. In 1960, hydroelectric plants produced 145 billion kilowatt-hours (19 percent of the total) (Edison Electric, 1977) and in 1976 283 billion kilowatt-hours (about 14 percent of the total) (U.S. Department of Energy, 1978a). Hydroelectric power's share of total electrical capacity has declined to the extent that even nuclear capacity has exceeded it.

The Federal Power Commission (1968) (FPC) estimated the potential conventional hydroelectric capacity of the entire United States to be about 180 gigawatts, but the constraints of available water resources and the difficulty of siting new dam and pumped storage units are likely to lower the achievable capacity considerably. Therefore, the Federal Power Commission (1976) estimates 1990 capacity at only 82 gigawatts, and hydroelectric generation at 300 billion kilowatt-hours by 1990--only 5 percent above the 1976 level. The Supply and Delivery Panel figure is slightly higher than the FPC estimate, but not significantly so. For the year 2010 the panel's figures saturate at about 500 billion kilowatt-hours (equivalent to 5 quads of input energy) in all three scenarios. For all scenarios, hydroelectric power, although important, is projected to decline in terms of total generated electricity and will not alter the need to convert to coal and nuclear energy as the primary means of base-load generation.

POTENTIAL NEW ENERGY SOURCES FOR THE GENERATION OF ELECTRICITY

Until the end of this century, the impact of advanced sources of electrical energy (solar thermal, ocean thermal, wind, photovoltaic, geothermal, and fusion) will be negligible, although several advanced technologies offer significant potential for the more distant future. As most of these sources require major development, it is not possible to predict when and how much they might be used, but they are not expected to provide more than a few percent of electricity generation before the turn of the century. For example, assuming technical success in developing geothermal and solar electric technologies, the enhanced supply scenario would show a little more than 3 quads of input energy for electricity generation (see Chapter 6).

NUCLEAR/COAL COST COMPARISON

Economic comparisons of nuclear and coal-fired power plants are subject to many variables. For example, geographic variations in construction costs, or transportation costs that significantly affect the delivered cost of coal, are major factors in comparing these two sources. The purpose of this section is to provide a general view of these comparative economics on a representative basis, assuming that current licensing and pollution control standards will remain stable.

Capital Costs

The capital costs of base-load generating facilities account for a large fraction of the total costs of generation. The following is an attempt to derive the basic cost comparison between nuclear plants and coal plants by analyzing the actual and forecast costs of existing projects, comparing these with analytical forecasts of several experts in the field.

Data made available through an informal association of electric power companies reflects actual experience in constructing nuclear and coal-fired generating units in various regions, at different times, using different labor forces and a wide assortment of architect-engineer and construction management firms and equipment suppliers. In compiling these data, all costs were reported on a common basis, so that overall estimates—and the elements that comprise them—can be directly and validly compared, with the resulting pooled data as a representative data base for projecting future costs.

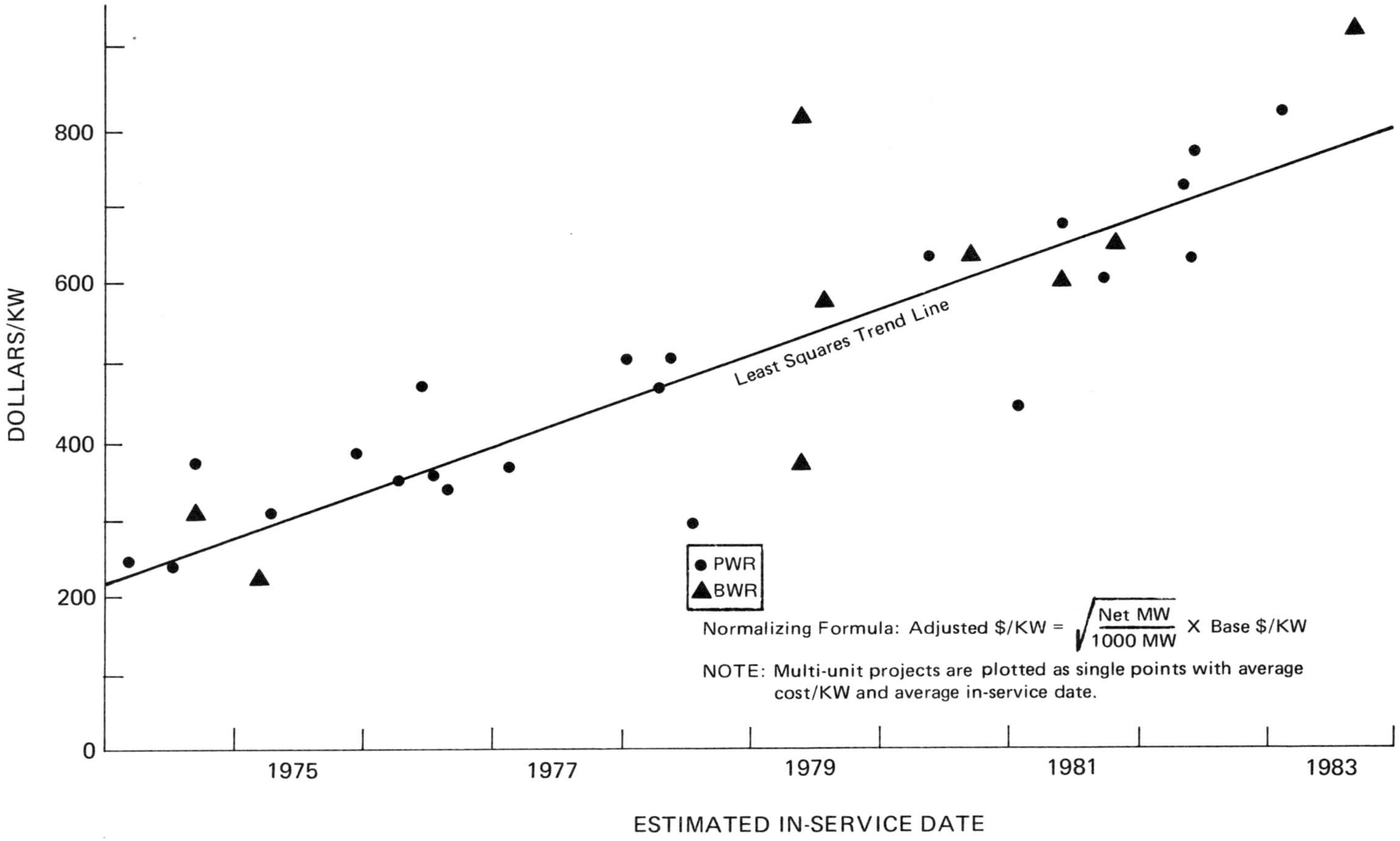

Figure 11 Comparative capital costs as a function of in-service date for nuclear generating stations from 1974 to 1983, in dollars per kilowatt of capacity (normalized to 1,000 megawatts). Data from pooled utility estimates for actual projects.

Figure 11 is a graphic summary of the pooled data on nuclear generating unit costs. This figure plots the actual or projected costs of 31 nuclear projects, with in-service dates spanning the 10-year period from 1974 through 1983. Costs are presented on a basis of dollars per kilowatt of capacity, with unit capacity normalized to 1,000 megawatts using the formula shown on the figure. Allowance for Funds During Construction (AFC)--interest and dividends on money used to finance construction--is not included in the costs in Figure 11.

The statistical average of the costs presented in Figure 11 is defined by the solid line, which has been calculated by the least-squares method. The slope of the least-squares trend line is about $60 per kilowatt per year, or effectively about 14 percent per year through the 10-year period depicted.

It should be noted that this rather high rate of increase does not necessarily reflect a simple cost escalation. It is a composite annual rate of increase in the cost of constructing new nuclear generating units, which has been driven not only by the higher cost of money but also by increasingly complex licensing procedures and requirements, new environmental requirements, recession-related delays, the mistakes common to any learning experience, and so forth.

The data point spread relative to the least-squares trend line in effect represents the range of uncertainty in the execution of large power plant projects, along with geographic differences and other variable factors. Mathematically this is expressed as the "standard deviation," which for these data was computed to be $93.3 per kilowatt. In other words, this data indicates that although the average capital cost of a nuclear unit lies along the least-squares line, there are substantial cost variations from unit to unit. For example, a 1,000-megawatt nuclear unit to be placed in service on January 1, 1984, would have a probability of about 70 percent of costing between about $700 and $900 per kilowatt (without AFC).

In Figure 12 similar cost data are presented for 14 coal-fired projects. For the purpose of this example, unit size is normalized to 800 megawatts, a representative size for new coal units. The range of in-service dates spans the period from 1974 through 1984, and AFC is again not included. In all cases, the costs of the coal-fired units presented on this figure include all design features necessary for the plants to be in full compliance with federal standards for stack emissions applicable in 1977.

Coal data have been analyzed in a manner similar to that of the nuclear data. The least-squares trend line as shown has a slope of about $45 per kilowatt per year or effectively 16 percent per year, over the same 10-year period. The standard deviation is calculated to be $78 per kilowatt. The apparent difference in capital cost growth rate between fossil and nuclear plants is probably an artifact of the data and of no fundamental significance. Coal and nuclear plants have many of the same problems; although fossil plants do not have nuclear licensing costs, they have been affected by strict air and water pollution standards.

Figures 11 and 12 indicate that the installed cost of a typical nuclear or coal-fired generating unit to be placed in service in 1983-

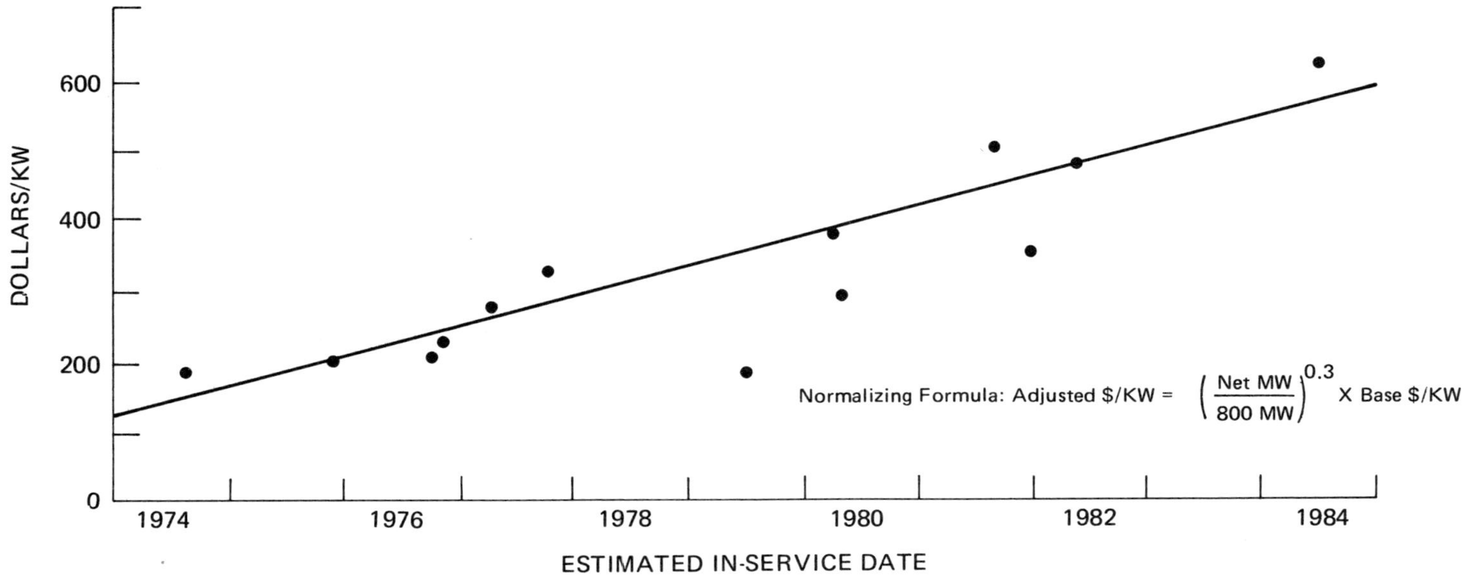

Figure 12 Comparative capital cost as a function of in-service date for coal-fired generating stations from 1974 to 1983, in dollars per kilowatt of capacity (normalized to 800 megawatts). Data from pooled utility estimates for actual projects.

Normalizing Formula: Adjusted $/KW = $\left(\frac{\text{Net MW}}{800 \text{ MW}}\right)^{0.3}$ X Base $/KW

84 will be about $800 per kilowatt and $550 per kilowatt respectively, without AFC. To provide some measure of corroboration of these estimates, we have compared them to similar but independent projections published by recognized experts in the field of generation economics; the results are summarized in Figure 13 (with AFC).

Figure 13 presents, in bar-chart form, the installed costs of nuclear and coal-fired generating units, as estimated in separate studies. Each of these studies predicts unit costs spanning the period 1983 to 1987. The figure uses adjusted source data where adjustments were required to permit direct comparison of 1983-84 costs and equivalent unit sizes. These adjustments were made using the trend lines of Figures 11 and 12.

Since the capital cost estimates of each of these four studies include AFC, direct comparison with the capital cost estimates of Figures 11 and 12 (both of which exclude AFC) requires some adjustment. Nuclear plant AFC represents about 20 percent of the total project cost, and for coal-fired plants AFC is approximately 16 percent of the total project costs. The cross-hatched bars on Figure 13 show the corresponding upward adjustment of pooled utility data.

In conclusion, based on statistical analysis of multiple, independent power plant cost projections, compiled using a pooled data base, the panel estimates that the capital cost of a 1,000-megawatt nuclear generating unit to be placed in service in 1983-84 will be $800 per kilowatt without AFC, or $950 per kilowatt with AFC. Although the analysis is based on 1,000 megawatts, a more typical nuclear plant would be 1,200 megawatts with a corresponding 1983-84 cost with AFC of about $850 per kilowatt. The indicated capital cost of an 800-megawatt coal-fired unit, meeting current air and water pollution standards, would be $550 and $650 per kilowatt without and with AFC, respectively. The ratio of coal-to-nuclear unit costs will be from 75 to 80 percent.

Electric utilities recognize that nuclear plants are more expensive to build than fossil plants, simply because they are more sophisticated in design and require more concrete, steel, circuitry, and so on. Despite changing requirements for nuclear and coal plant licensing, this fundamental difference will continue to exist. Moreover, it is probable that certain unknowns (such as the cost of money, material availabilities, and labor rates) will influence construction costs in years to come and will have roughly equivalent impacts on both nuclear and fossil projects. Therefore, the capital-cost ratio of coal and nuclear generating units is more predictable and more precise than absolute cost estimates and, as a consequence, is useful as a basis for economic comparisons of coal-fired and nuclear capacity.

Generation Costs: Fossil and Nuclear

Starting with the capital costs identified above, Table 13 summarizes the busbar cost of generation from a new 1,200-megawatt nuclear plant and a new 800-megawatt coal plant to begin operating in 1983-84. Assuming plant operation of 7,000 hours per year, fixed charges are estimated using a fixed charge rate of 18 percent. The fixed charge rate

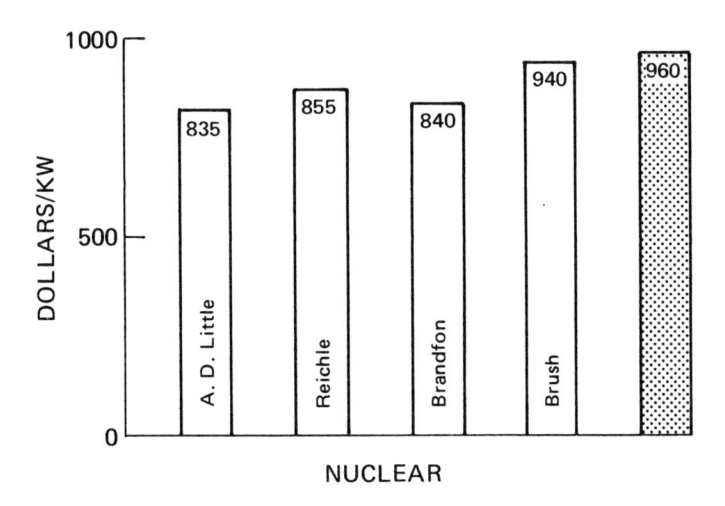

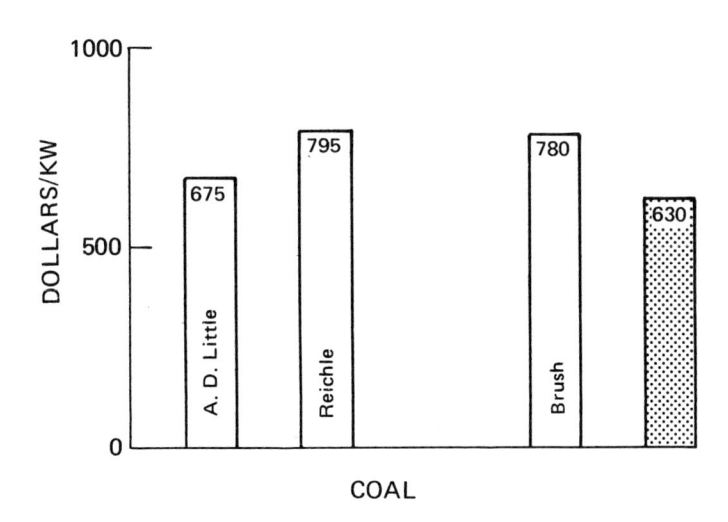

Figure 13 Projections (unshaded bars) of installed costs for nuclear and coal-fired generating units to be put in service in January 1984, and pooled utility capital cost data from coal-fired and nuclear plants represented in figures 11 and 12 (shaded bars), in dollars per kilowatt of capacity. Projections from Arthur D. Little (1975), Reichle (1975), Brandfon (1975), and Brush (1976).

is approximately the rate necessary to provide revenues to cover financing costs, income taxes, and depreciation, and depends on the cost of money and variable state regulatory practices. The 18-percent rate is chosen for illustrative purposes. The fuel costs are chosen to reflect recent price levels for uranium and Appalachian coal. The cost of coal, of course, will vary greatly with transportation requirements. The operating and maintenance cost (minus the fuel cost) of the coal plant is approximately doubled by the estimated cost of operating the sulfur dioxide removal system.

Table 13 indicates that a typical nuclear plant will be about 10 to 15 percent less expensive on the basis of a first 10-year average. The greater variable cost subject to inflation in the case of the coal plant will result in an increased advantage for nuclear power when viewed over the life span of 30 years. Similarly, the greater fixed charges of the nuclear plant will decrease the nuclear advantage if less than 7,000 hours of generation per year are achieved.

Variations in economic comparisons from one geographic region to the next will result in comparable variations in commitment to coal and nuclear generation. Beyond simple economics, each utility system is motivated to attempt to diversify its energy sources to reduce the vulnerability of the power supply to any one technological or political event, such as a coal strike.

Table 13 Average generation costs first 10 years with startup in 1983/1984, in mills per kilowatt-hour[a]

Cost component	Nuclear ($800/kw)	Coal ($600/kw)
Capital	24	19
Fuel	12	20
Operation and maintenance	4	6
Total	40	45

[a]Basis: 18-percent fixed charge rate; 7000 hours per year; 1976 fuel—U_3O_8 at 40 dollars per pound, coal at 24 dollars per ton; 5-percent inflation.

Construction schedules and preconstruction lead times for all power plants have lengthened significantly during the last 10 years. The total lead time for a nuclear plant is a minimum of 9 years, whereas it takes about 7 years to get a coal-fired plant on line. Protracted lead times have been the result of requirements for various permits prior to construction. In addition, construction times have been lengthened by the adverse impact of plant complexity on labor productivity, as well as the equipment and material needed to meet ever more stringent safety and environmental requirements.

Long construction times coupled with sharp inflation and high financing costs have contributed greatly to capital increases of large power plants. It has been estimated that the final capital cost is increased by a factor of two by the combined effects of inflation and financing costs. In addition, the high cost and long construction times have influenced the financial health of the utility industry by raising the need to provide large sums for construction financing; these sums, in most regulatory jurisdictions, are not recognized in determining utility revenue requirements until after the plants are in service.

Beyond financial factors, long lead times naturally place strong emphasis on long-range load forecasting. If the forecast is in error or if customers behave in a manner different from that assumed, it is virtually impossible to respond in a way that does not result in drastic reductions in generating reserves on one hand or drastic increases in financing costs for deferring operation of unneeded capacity on the other. Despite this, a recent study (Decision Focus, 1978) indicates that the total social costs of under-capacity are substantially greater that those of over-capacity.

NONTECHNICAL FACTORS

Nontechnical factors affecting electricity generation and transmission can be divided into those that act on the generation and distribution facilities themselves and those related to the primary energy sources used to generate electricity. This section will discuss only the former, and constraints applying to the fuels will be discussed in the chapters devoted to those fuels.

Electric systems of all types face ever-increasing difficulties in meeting legal, regulatory, and administrative requirements. Electric power plants and transmission lines have become a political focal point for action in responding to the national concern with protecting the environment. At the same time, despite these environmental concerns, public consumption of electricity continues to increase. The problem is to place these apparently conflicting concerns in better balance and perspective.

In this connection, the Federal Power Commission (1976) reviewed the experience with two hydroelectric projects, two nuclear power plants, one coal-fired plant, and one transmission line to illustrate the difficulty of meeting electric energy needs in the face of the multiple, complex, and overlapping federal, state, and local requirements. Of the six examples, the two hydroelectric projects and the coal-fired plant have been rejected or abandoned, and the other three projects experienced serious delays at great increases in cost to consumers. The seriousness of the problem is illustrated by the fact that to replace the electric power that would have been generated by the abandoned projects with power to operate the oil-fired plants would require about 70 million barrels of oil per year (Federal Power, 1976). This amounts to more than 3 percent of the oil imported by the United States in 1976.

There are many examples of power plants and transmission projects that have been rejected or abandoned with consequences even more severe than those of the aforementioned examples. Such difficulties lead to higher costs and problems of financing new plants to meet future needs, causing many analysts to see financing as one of the major hurdles for most of the nation's electric utilities.

In a recent survey of utilities with plans for nuclear power stations, the utilities expressed their concern about problems that increase in number and severity and tax their ability to develop sites and build nuclear power stations of any size (Cope and Bauman, 1977). Their concerns were much broader than just the siting of nuclear power stations; they cited numerous cases of difficulty in obtaining approval for siting transmission and distribution facilities.

In summary, the most serious constraints on the generation and distribution of electricity are those related to the siting and construction of generation and transmission facilities, with the resulting financial problems posing an equally serious potential future constraint. At this time, generating sites and facilities are suffering extreme delays, often are abandoned, and always incur major cost overruns, which add to the cost of electricity to the consumer.

Electric Utility Decision Making

This chapter has outlined the difficulty of determining which of the several generation alternatives available to utility managements best permit them to meet their franchised obligations to provide safe, reliable, and economical electric service. A multitude of tangible and intangible factors must be considered to permit selection of the generating alternative likely to be best suited over its 30- to 40-year lifetime to the needs of a particular utility.

One way to view this decision-making process is to look at the net effect of all prior decisions and the effect on the mix of fuels. Figure 14 illustrates the fuel mix of electric generation for different areas of the United States. It is obvious that different parts of the country use vastly different proportions of available fuels to provide their electric generation. For example, in the Northwest, where hydroelectric sites are relatively abundant, fossil fuels meet only a small portion of the demand for power. In the Southwest, where much of the nation's oil and natural gas is extracted, natural gas is the predominant fuel for generation.

These past decisions were often determined by past conditions. The world has seen recent major changes in its energy regime, however, and without doubt these changes will continue. One of the major challenges facing electric utility managements is to evaluate these future changes properly over the next 7 to 10 years in order to make appropriate decisions today on generating units that will not go into service for almost a decade.

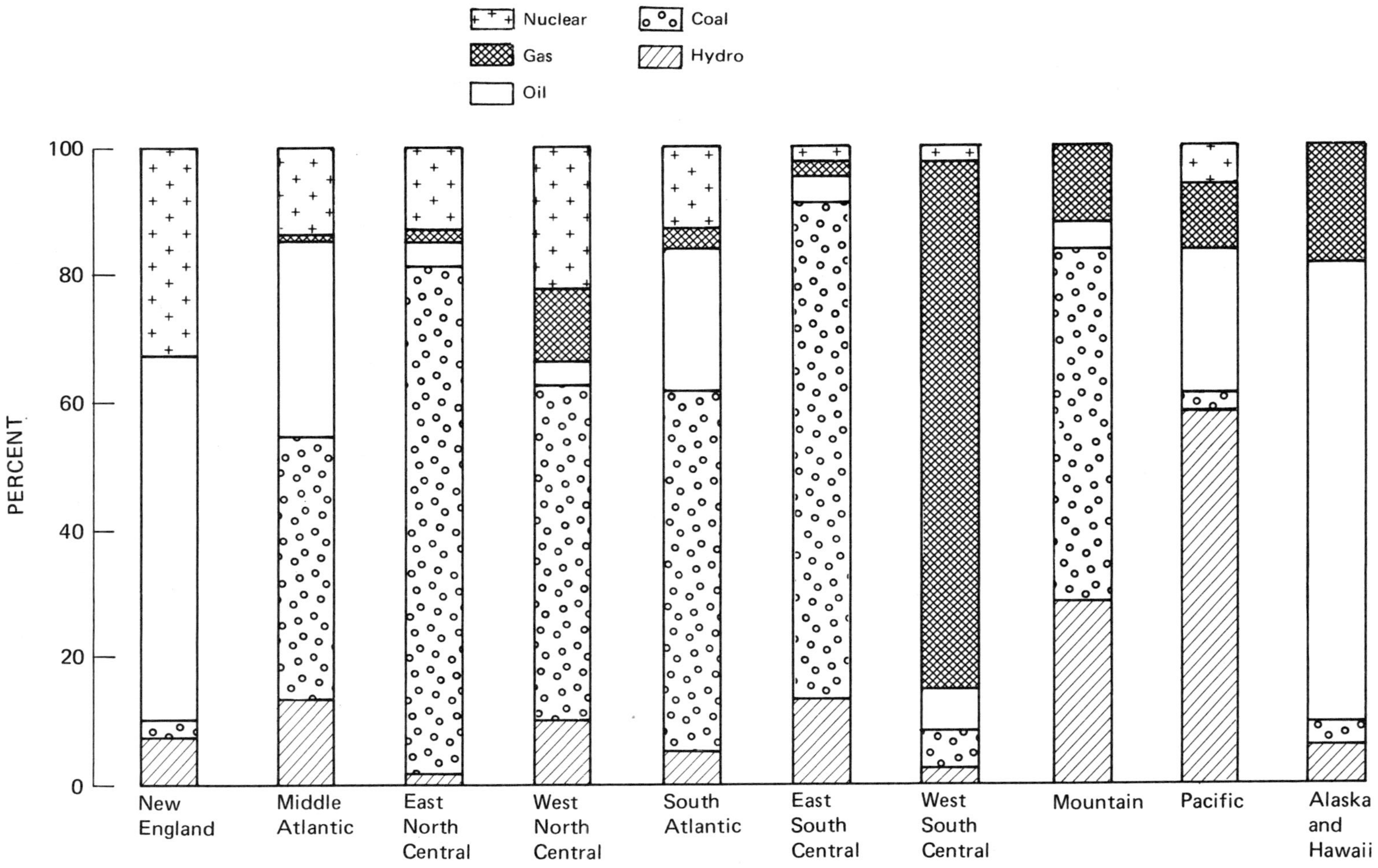

Figure 14 Fuel mix for electricity generation for different regions of the United States, by percentage of regional total.

The first step in meeting this challenge is determining that a need exists. As was noted earlier, experts in load forecasting cannot agree on what reasonable long-range rates of growth the utility industry might anticipate. Yet a decision must be made, because there is general agreement that there will be growth at some rate. Without a decision, reliability will be degraded and utility managements will not have met their franchised obligation to provide adequate reliability.

One factor assuming greater importance in decision-making is fuel supply reliability. The Arab oil embargo and the heavy reliance of the United States on imported foreign oil--when combined with depletion of domestic oil reserves--illustrate this point. The rate of depletion of our natural gas resources, which may require conversion of many gas-fired electric generators to other fossil fuels, is another instance of changing reliability of fuel supply. The protracted coal strike of the winter of 1977-78 illustrates that even abundant domestic fuel resources can be subject to long-term interruptions. Environmental policies also can affect fuel source reliability. For example, in the late 1960s, air pollution regulations imposed along the East Coast forced conversion of many electric generating stations from coal to low-sulfur oil. With the problems of fuel reliability assuming more and more significance, electric utility managements have had to both diversify their sources of supply for a given fuel and critically evaluate the fuel mix of their complete generation systems. The concept of not putting all one's eggs in one basket, thereby permitting greater flexibility to meet changing conditions, has assumed far greater importance.

In addition to reliability, utilities are charged with providing safe and economical service. At one time safety meant just electric safety, but recently it has come to connote environmental safety in its broadest sense. Environmental safety runs the gamut from human safety in nuclear power generation and air pollution control all the way to ecological safety with respect to water pollution control and land use. Resolutions of the problems arising from the new national awareness of environmental protection affect and will continue to affect the obligations of utility managements, strongly influencing decision making.

The final phase in the decision-making process is evaluating the economic alternatives available considering the need for reliability and environmental safety. The selection of plant type depends on the type and magnitude of the load. Short-duration peak loads can be most economically met with low-capital-cost plants, which have inherently higher fuel costs since they burn natural gas and oil. On the other hand, burning cheaper fuels in an environmentally acceptable manner calls for more capital investment per unit of electrical output and, therefore, requires the maximum possible use to be economical.

Figure 15 shows a typical electric utility load duration curve. Imposed on it are the relative total costs per kilowatt per year to produce energy from different types of generation depending on its output duration. From the load duration curve, which shows the average load as 60 percent of the peak load and further indicates the relatively short period of time a peak exists (approximately 1,500 hours

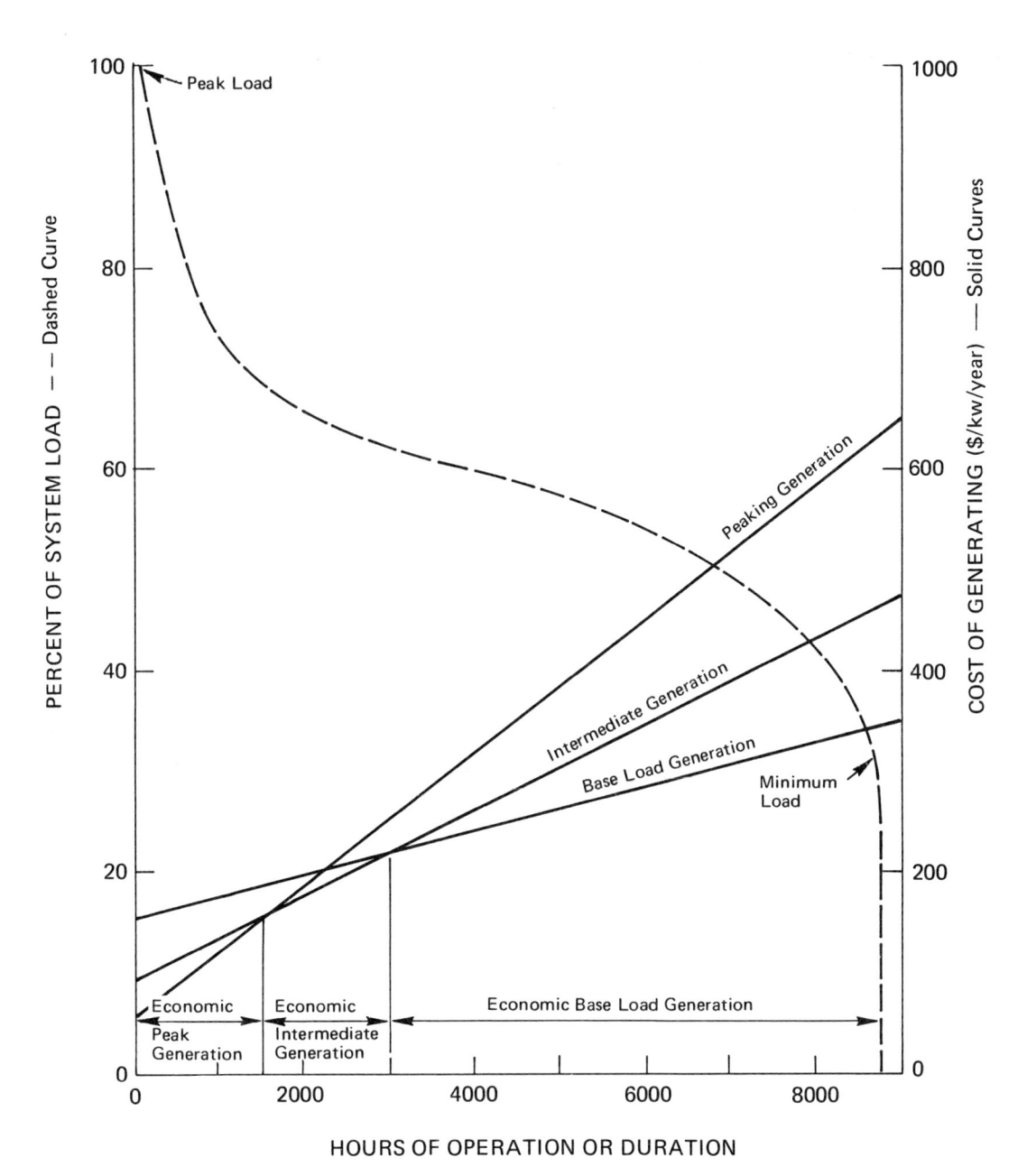

Figure 15 Utility load duration curve (dashed curve), by percentage of system peak load, and generation costs for base load, intermediate and peaking generation (solid curves), showing dependence of costs on load duration.

per year), we can see the need both for generation that is economical when operating only a few hours out of the year to meet peaks, and base-load generation required to operate continuously. The typical cost shown for the three modes of generation--peak, intermediate, and base--demonstrate that maximum economy can be obtained by using the peaking units (with their low fixed costs but high operating costs) a few hours each year, using intermediate-load units 1,500 to 3,000 hours per year, and using base-load generation (with its high fixed costs but low operating costs) for operation in excess of 3,000 hours per year.

In summary then, the decision-making process must properly weigh the corporate obligation to provide safe, reliable, and economic electric service consistent with needs and desires of society spanning half a century. Meeting this commitment in today's world of rapid social and political change is a major challenge to utility managements.

REFERENCES

Arthur D. Little/S. M. Stroller. 1975. Economic Comparison of Base-Load Generation Alternatives for New England Electric. Westborough, Mass.: New England Electric.

Ashton, W. 1977. Wind Power. West Virginia University Magazine 9(2):2-5.

Atomic Industrial Forum. 1977. Nuclear Energy Production Takes Big Jump in 1976, Offsets 325 Million Barrels of Oil, Saves 1.4 Billion. INFO News Release. Washington, D.C.: Atomic Industrial Forum, April 27.

Averitt, Paul. 1975. Coal Resources of the United States, January 1, 1974. Bulletin 1412, Geological Survey, U.S. Department of the Interior. Washington, D.C.: U.S. Government Printing Office (024-001-02703-8).

Brandfon, William W. 1975. Economics of Nuclear Power. Paper presented at Atomic Industrial Forum Conference on Nuclear Power Financial Considerations, Bal Harbour, Fla., December 8. Available from William W. Brandfon, Sargent & Lundy, Inc., 55 East Monroe St., Chicago, Ill. 60603.

Brush, Harvey F. 1976. Power Plant Economics. Testimony presented at Connecticut Public Utilities Control Authority, Hartford, Conn., January 21. Available from Harvey F. Brush, Bechtel, Inc., Box 3965, San Francisco, Calif. 94119.

Cope, D. F., and H. F. Bauman. 1977. Utility Survey on Nuclear Power Plant Siting and Nuclear Energy Centers. Oak Ridge, Tenn.: Oak Ridge National Laboratory (ORNL/TM-5928).

Decision Focus. 1978. Costs and Benefits of Over/Under Capacity in Electric Power System Planning. Prepared for Electric Power Research Institute. Palo Alto, Calif.: Electric Power Research Institute (EA-927).

Ebasco Services. 1977. 1977 Business and Economic Charts. New York: Ebasco Services.

Edison Electric Institute. 1976a. Economic Growth in the Future. New York: Edison Electric Institute.

Edison Electric Institute. 1976b. Statistical Yearbook for 1975. New York: Edison Electric Institute.

Edison Electric Institute. 1977. Statistical Yearbook for 1976. New York: Edison Electric Institute.

Electrical World. 1977. 28th Annual Electrical Industry Forecast. Electrical World 188(6):43-58.

Electricity Conversion, Transmission, and Storage Resource Group. 1977. Report to the Supply and Delivery Panel, Committee on Nuclear and Alternative Energy Systems, National Research Council, Washington, D.C.

Federal Energy Administration. 1974. Project Independence. Washington, D.C.: U.S. Government Printing Office.

Federal Power Commission. 1965. Hydroelectric Power Resources of the United States: 1964. Washington, D.C.: Federal Power Commission. Available for inspection or photocopying at the U.S. Department of Energy, Division of Energy Information, 825 N. Capitol St., Washington, D.C. 20426

Federal Power Commission. 1968. Hydroelectric Power Evaluation. Washington, D.C.: Federal Power Commission (FPC P-35). For availability see Federal Power Commission (1965).

Federal Power Commission. 1971. The 1970 National Power Survey, Part 1. Washington, D.C.: Federal Power Commission. For availability see Federal Power Commission (1965).

Federal Power Commission. 1976. Factors Affecting the Electric Power Supply, 1980-85. 4 vols. Washington, D.C.: Federal Power Commission. For availability see Federal Power Commission (1965).

Institute for Energy Analysis. 1976. Economic and Environmental Implications of a U.S. Nuclear Moratorium, 1985-2010. Oak Ridge Associated Universities. Oak Ridge, Tenn.: Oak Ridge National Laboratory (ORAU/IEA 76-4).

Kidder, Peabody & Co. 1978. Nuclear Reactors in the United States as of December 31, 1977. Research Department. Washington, D.C.: Kidder, Peabody & Co.

Lester, Richard K. 1978. Nuclear Power Plant Lead-Times. Working Paper, International Consultative Group on Nuclear Energy. New York: The Rockefeller Foundation/The Royal Institute of International Affairs.

Lilienthal, David E. 1977. Lost Megawatts Flow Over the Nation's Myriad Spillways. Smithsonian Magazine 8(6):83-89.

Morgan Guaranty Trust. 1978. When Will the Lights Go Out? Morgan Guaranty Survey. New York: Morgan Guaranty Trust.

National Coal Association. 1972. Coal Data. Washington, D.C.: National Coal Association.

National Electric Reliability Council. 1977. Fossil and Nuclear Fuel for Electric Utility Generation: Requirements and Constraints, 1976-1986. Princeton, N.J.: National Electric Reliability Council.

National Research Council. 1978. Energy Modeling for an Uncertain Future. Modeling Resource Group, Synthesis Panel, Committee on Nuclear and Alternative Energy Systems. Supporting Paper 2. Washington, D.C.: National Academy of Sciences.

National Research Council. 1979. Alternative Energy Demand Futures. Demand and Conservation Panel, Committee on Nuclear and Alternative Energy Systems. Washington, D.C.: National Academy of Sciences.

Pollard, William G. 1976. The Long Range Prospects for Solar Energy. American Scientist 64:424-429.

Reichle, Leonard. 1975. Economics of Nuclear Power. Paper presented to the New York Society of Security Analysts, New York, August 27. Available from Ebasco Services, Inc., 2 Rector St., New York, N.Y. 10006.

U.S. Department of Energy. 1978a. Annual Report to Congress, 1978: Vol. 2, Data. 3 vols. Energy Information Administration. Washington, D.C.: U.S. Department of Energy (DOE/EIA-0173/2).

U.S. Department of Energy. 1979. Monthly Energy Review: June Energy Information Administration. Washington, D.C.: U.S. Department of Energy (DOE/EIA-0035/6).

U.S. Department of the Interior. 1975. United States--Energy through the Year 2000 (Revised). Bureau of Mines. Washington, D.C.: U.S. Department of the Interior (76B002202A).

3 OIL AND GAS

Oil and natural gas are the preferred fuels worldwide, accounting for 75 percent of domestic energy consumption and about 70 percent of the world's energy consumption. They are indispensable for American transportation and are important but less critical for residential, commercial, and industrial uses. Petroleum and natural gas now fuel about 30 percent of the nation's electrical generation capacity. This chapter discusses oil and gas supply in its broadest sense including not only petroleum and natural gas, but also shale oil and synthetic oil and gas from coal.

Domestic oil and natural gas production peaked in the early 1970s and is now declining. This decline, combined with growing demand, fosters increasing U.S. dependence on imports, particularly of crude oil. World oil production is expected to peak in turn near the end of this century, and unless demand is moderated significantly, U.S. import requirements will be met--if at all--only at greatly increased costs and with increasingly unacceptable political risks.

PETROLEUM

Estimates of the future availability of petroleum (crude oil and natural gas liquids) have been numerous and varied. The Oil and Gas Resource Group of this panel appraised several of these estimates, and selected those that seemed most reliable (Table 14). The total amounts are customarily divided into two categories: recoverable reserves and recoverable potential resources. Recoverable reserves are deposits that have been identified that can be extracted using available technology at prevailing prices. Recoverable potential resources are deposits, believed

Table 14 U.S. and world recoverable reserves and estimated recoverable potential resources of crude oil and natural gas liquids as of December 31, 1976

Location	Crude oil			Natural gas liquids			Total		
	Recoverable reserves	Recoverable potential resources	Total	Recoverable reserves	Recoverable potential resources	Total	Recoverable reserves	Recoverable potential resources	Total
	Billions of barrels								
U.S.	31.3	76	107	6.4	10	16	37.7	86	124
Other market economies	510	537	1047	66	70	136	576	607	1183
Centrally planned economies	101	350	451	13	45	58	114	395	509
World total	642.3	963	1605	85.4	125	210	727.7	1088	1815
	Quads (10^{15} Btu)								
U.S.	175	426	601	26	40	66	201	466	667
Other market economies	2856	3007	5863	266	282	548	3122	3290	6412
Centrally planned economies	566	1960	2526	53	181	234	619	2141	2760
World total	3597	5393	8990	345	503	848	3942	5897	9839

Source: Moody (1975) for recoverable potential resources; reserve estimates from Supply and Delivery Panel resource group reports (available in CONAES public file).

to exist, that might eventually be economically feasible to extract. These definitions obviously depend on the economics of production. Stores of petroleum will move from resources to reserves as discoveries are made or as prices, policies, and technologies make the deposits economically recoverable. The United States, which consumes more than one-fourth of the world's petroleum production, contains only about 5 percent of the world's recoverable reserves.

A reserve-to-production ratio of about 10 to 1 is considered necessary to ensure a steady supply of oil. Thus, additions to reserves must keep pace with production if production is not to decline. In the United States, production has outpaced reserve additions since about 1968. Production itself peaked in about 1970 and has since declined. Maintaining domestic reserves at levels high enough to sustain current annual production (about 20 quads) is unlikely, because additions to reserves now must come from generally less accessible locations and at correspondingly high costs. Thus, even holding production constant will require continually increasing the effort and expense devoted to exploration and production.

A measure of this difficulty is the finding rate, or the amount of oil added to reserves per foot of exploratory and development drilling. Although this rate fluctuates (for example, when a large deposit like that at Prudhoe Bay is discovered), it has declined steadily during the last decade. This panel believes that large discoveries in the continental United States are unlikely. Thus, if the finding rate is to be maintained or improved, new technologies and improvements on existing ones are needed. With current techniques, only about a third of the oil in a typical reservoir can be recovered. Technologic needs include improved knowledge of plate tectonics, a better understanding of geothermal gradients and their influences on oil generation, and improved remote sensing techniques. The greatest need, of course, is for a direct finding method. Whether or not technologies are improved, it is certain that increased drilling will be necessary to maintain production near current levels and that such drilling will become increasingly costly.

U. S. Domestic Production

Annual crude oil production in the United States peaked in 1970 at 3,517 million barrels; natural gas liquids production peaked in 1972 at 638 million barrels (total in 1970 of 4.123 million barrels). By 1975, production of crude oil and natural gas liquids had dropped more than 10 percent to 3,650 million barrels. Between 1972 and 1975, annual imports of crude oil, natural gas liquids, and refined oil products increased from 1,735 million to 2,176 million barrels (U.S. Department of the Interior, 1976a). The decline of domestic production has been halted temporarily by the oil produced in the Alaskan North Slope but this new production, which is expected to reach 440 million barrels per year in 1979, will compensate for the continuing decline of older resource basins for only a year or two.

The production rate for domestic oil is influenced by the demand for its use, cost-price and profit considerations, the availability of deposits, and the general financial and regulatory climate. Steadily increasing domestic consumption attests to the demand for oil. This panel concluded that the policies and regulatory practices of the government and the resulting business climate were likely to influence the producibility of oil more than the price at which it is sold. Accordingly, it did not attempt to derive a specific price-production curve, projecting instead what might be produced under different hypothetical assumptions about national policies and practices. Three "scenarios" were developed to describe production under different sets of assumptions: labeled business-as-usual, enhanced supply, and national commitment. Tables 15 and 16 give the panel's estimates of the effects of these various conditions.

The business-as-usual scenario was derived by assuming little or no change in government and industrial policies concerning exploration for and production of domestic oil. Under this scenario, prices would continue to be held by federal control below world market prices; current environmental requirements, including lengthy impact statements and adversary hearings, would be retained; public lands would continue to be withdrawn from exploration and production; and outer continental shelf development would continue to require separate permitting processes for exploration and production. Government energy policies would remain inconsistent, and little effort would be made to encourage the raising of capital funds for oil exploration and production.

Under an enhanced supply scenario, it is assumed that some barriers to increased production would be removed. Federal offshore leasing would be accelerated, wellhead oil and gas prices would be decontrolled, and exploration and production technology would evolve, spurred largely by higher prices. No changes would occur in onshore land leasing or in environmental controls, except for simplification of the permit process.

The national commitment scenario depends on the basic assumption that government and industry would cooperate to produce as much oil as possible. Stipulations of the Clean Air Act would be relaxed; preparation and review of environmental impact statements would be simplified and speeded up, although environmental standards themselves would remain unchanged; loan guarantees and other incentives would be increased to promote development by industry of technology for oil exploration and production; some federally withdrawn lands would be made available for exploration and production; the industry would be assured federal priorities on goods, services, and labor; and tertiary methods to recover additional oil would become economical.

In deriving the expected production levels under these sets of assumptions, it is necessary also to estimate what might happen to reserves and resources, since these are also influenced by government and industrial policies and practices. It was estimated that with the enhanced supply assumptions, increased drilling and the resulting additional finds would expand both reserves and resources about 10 percent over current levels. In a national commitment scenario, not only would an increased finding rate result, but application of tertiary recovery methods at existing oil fields might significantly increase the percentage of oil recovered from reservoirs, creating additional reserves.

Table 15 U.S. petroleum reserves and resources

Scenario	Recoverable reserves		Recoverable potential resources		Total recoverable resources	
	(10^9 bbl)	(quads)	(10^9 bbl)	(quads)	(10^9 bbl)	(quads)
Business as usual	38	201	86	466	124	667
Enhanced supply	42	222	94	509	136	731
National commitment	48	254	107	580	155	834

Table 16 U.S. petroleum production to 2010, in quads per year

Scenario	1975	1985	1990	2000	2010
Business as usual	20	18	16	12	6
Enhanced supply	20	21	20	18	16
National commitment	20	21	21	20	18

Table 16 illustrates that, according to these estimates, under present trends crude oil production will decline steadily from now through 2010, although effective national policies could hold domestic production reasonably constant. According to these estimates, there are no policies that will increase domestic production significantly.

World Production and Availability of Imports

Until World War II, the Untied States was self-sufficient in petroleum. After the war, it became a net importer of crude oil, because these early imports were less expensive than domestic oil. By 1965, petroleum consumption began to exceed domestic production capacity. By 1970, total imports were 23 percent of U.S. consumption; by 1975 they were 37 percent, and in 1977 they were almost 50 percent (U.S. Department of Energy, 1979).

Domestic production is declining while consumer demand grows and the nation's dependence on foreign sources continues to increase. Estimates from the previous section suggest that even with a national commitment to oil production, the United States will be able only to maintain present domestic production levels. There is little doubt that for the next 10 to 20 years substantial reliance on imports will be necessary.

Thus, the attention of the United States (and the world) has focused on the supply of petroleum. There is concern that before the end of this century petroleum exporting countries will become unable or unwilling to produce the amounts of oil required by consuming nations (Rustow, 1977; Levy, 1977). The gap between U.S. demand and supply will have to be met by some other means. Among these other means are extraordinary increases in world oil prices (serving to choke off demand), a slowing of economic growth in importing countries, and substitution of alternative sources of energy for oil.

The Petroleum Resource Group of this study examined the outlook for oil imports. It estimated the probable oil output of the petroleum exporting countries under various conditions, ranging from a highly pessimistic example in which the countries need to produce near their physical capacities to meet demand, to an optimistic one in which they are assumed able to lower production (Table 17). Estimates for the noncommunist countries are based on discussions with industry and foreign experts. In most cases, they are little more than informed guesses. The estimates for Saudi Arabian production in this discussion are exceptions, being based not on what the Saudis can or will produce, but rather on how much they would have to produce to reach an assumed total free-world demand. (The total demand estimates used in Table 17 agree substantially with those of the recent Workshop on Alternative Energy Strategies [1977] study.)

In reviewing the production potential for world crude oil, the Petroleum Resource Group concluded that growth in world oil production would probably be concentrated in a relatively few countries. About a third of the increase between 1975 and 1995 would come from the Communist countries, principally the Soviet Union and the People's Republic of China. Another third would occur in Arab countries, with Saudi Arabia alone supplying 22 percent of the increase. The remaining third would be divided among other countries, with the greatest increases in Western Europe and Mexico. It was assumed that the Soviet bloc would be self-sufficient in oil and, as a whole, would increase production only to satisfy its own demand. It is assumed that China would not export significant amounts of oil.

However, most if not all noncommunist producers, including the United States, will probably operate below full capacity as a matter of design and circumstance. For example, Venezuela and Kuwait are already restricting output to conserve what appear to be depleting reserves.

Table 17 Estimated requirements for world oil production through 1995 under varying assumptions, in quads per year

	1975	1985	1995
Optismistic case			
Total world production	111.2	171.4	207.7
Total of market economies	87.8	131.0	153.6
OPEC production	54.7	81.2	90.3
Arab production	33.5	56.6	67.0
Saudi Arabian production[a]	13.7	27.9	35.1
OPEC share of market economies	62.3%	61.8%	58.4%
Arab share of market economies	38.2%	46.1%	43.4%
Saudi share of market economies	15.6%	21.3%	22.8%
Middle case			
Total world production		174.9	216.7
Total of market economies		134.5	162.6
OPEC production		88.9	106.7
Arab production		61.2	82.0
Saudi Arabian production[a]		33.1	50.9
OPEC share of market economies		66.1%	66.8%
Arab share of market economies		45.5%	50.4%
Saudi share of market economies		24.6%	31.3%
Pessimistic case			
Total world production		176.5	222.0
Total of market economies		136.1	167.9
OPEC production		95.0	124.2
Arab production		67.7	96.0
Saudi Arabian production[a]		40.2	66.5
OPEC share of market economies		70.0%	74.0%
Arab share of market economies		47.9%	64.2%
Saudi share of market economies		29.5%	39.6%

[a] Production assumed to equal the output needed to meet demand of market economies

Source: Johnson and Messick (1977)

For both countries, maximizing prices and restricting output make sense. It is virtually certain that Venezuela and Kuwait will further restrict their output as their conventional reserves of oil come closer to exhaustion. They will probably be joined by other producing countries, such as Iran, Algeria, and Indonesia, which by 1985 may all be in or approaching decline in crude oil production. Generally under such circumstances producing countries can be expected to try to preserve their reserve-to-production ratios by restricting output.

In addition, the future supply of oil will vary owing to political factors, military activity, technical difficulties, and other circumstances beyond the control of oil producers, such as the political situation in Iran and other oil producing nations.

Table 17 shows two other cases, which assume less optimistic production from all producing countries except Saudi Arabia and correspondingly increased Saudi production. The range of estimates about Saudi Arabia's needed production is from 35 to 65 quads per year to meet world totals of over 200 quads per year in 1995. This compares to current Saudi production of 13.7 quads per year. Saudi Arabia's ability to produce these amounts is doubtful, and its willingness even more so. The share of U.S. imports supplied by Saudi Arabia (and other OPEC nations) is rising steadily. This trend is not likely to change soon, and it may become a major problem. A reasonable estimate is that Saudi Arabia could probably produce 20 quads per year by 1980 and 40 quads per year after 1990. If so, the most optimistic estimate of 35 quads per year in 1995 might be possible, but the higher levels probably are not. If Saudi production were pushed to 50 quads per year by 1995, that level could be sustained for only 5 to 6 years and then would probably decline rapidly.

In fact, it is likely that, at some point, the Saudis will impose production limits well below their potential. It has been estimated that the Saudi economy could satisfy its foreign exchange needs with an output of only 3.5 million barrels per day, or 7 quads per year (Johnson and Messick, 1977).

In 1975, the United States consumed about 18 percent of the non-U.S. noncommunist oil output. If this share were maintained (which projections of world demand suggest is unlikely) imports could increase from the 1975 level of about 13 quads per year to a maximum of 22 quads per year in 1995, and then would start declining. That level is expected to be sufficient in the 1990's only if stringent conservation measures are taken, for after the 1990's much less oil would be available.

This discussion lends support to the need for strong action to reduce our reliance on oil imports, which has rapidly become a dependence on a single geographic area. It has been shown that oil imports are likely to be available only in limited amounts at increased costs by the 1990's, and after that, it is prudent to plan to reduce reliance on imported oil as rapidly as possible.

Oil Refining

To produce different types of liquid fuels, crude oil is processed in refineries. In the future, refineries will also process shale oil and synthetic crudes from coal. Domestic refineries process both domestic and imported crude oil. In addition, the United States imports finished oil products refined outside the country. Domestic refineries, on the average, now operate at over 90 percent of capacity. The average output from refineries is currently 48 percent gasoline, 22 percent distillate heating oil, 7 percent aviation jet fuels, 9 percent residual fuel oils, 3.5 percent petrochemical feed stocks, 3 percent petroleum coke, 3 percent asphalt, and the remainder specialty products such as solvents, lubricants, and waxes.

At the start of 1977, there were 276 domestic operating refineries (in 39 states) with a combined capacity of 16.7 million barrels of crude oil per calendar day (U.S. Department of the Interior, 1977). Individual capacities ranged from 10,000 to over 500,000 barrels per day (Oil and Gas Journal, 1977). From 1960 to 1965, refinery capacity increased at the rate of 1.25 percent per year, from 1965 to 1970 at 3 percent per year, and 1970 to 1976 at 4 percent per year. Uncertainties about future regulation of the oil industry have caused a sudden decline in additions to refining capacity. Only one domestic refinery (a 250,000 barrel-per-day project) is under construction; many new projects have been canceled or deferred, and additions for 1977, 1978, and 1979 are expected to be limited to strategic modifications of existing facilities to eliminate bottlenecks (Federal Energy, 1976).

The Petroleum Resource Group of the Supply and Delivery Panel of this study projected growth of U.S. refinery capacity between now and 2010 as shown in Table 18. Between 20 and 80 new 200,000 barrel-per-day refineries, including replacements and modernization capacity additions, will have to be built between now and 2010 to achieve the capacities shown.

Table 18 U.S. refinery capacity, in quads per year

Scenario	1977	1985	1990	2000	2010
Business as usual	33.4	37.0	37.6	40.0	40.8
Enhanced supply	33.4	40.4	42.0	47.8	52.8
National commitment	33.4	40.8	45.2	54.2	65.0

New off-gas and waste-water emission constraints will soon be placed on stationary emission sources, such as refineries and storage facilities.

Current emission regulations on sulfur and nitrogen oxides, hydrocarbons, particulates, and carbon monoxide will be tightened and possibly expanded to include other pollutants. Costs for environmental control equipment will increase rapidly as standards become tighter and more comprehensive. The net result will be cleaner operations from an environmental viewpoint, at the cost of substantial (probably 10 to 25 percent) increases in capital expenditures for refineries.

Another important consideration that inhibits the building of new refineries is the high inflation that has occurred over the last several years. Refiners have been reluctant to build new capacity that would operate at a loss in a market where prices are set by less expensive and fully depreciated old facilities.

The Oil Delivery System

Crude oil and refined oil products are transported by pipelines, tankers, barges, tank trucks, and railroad tank cars. In the United States most oil--except for heavier viscous products such as residual fuel oil--is transported through the nation's 220,000 miles of oil pipelines. About one-third of this mileage is made up of gathering lines from oil-producing areas, one-third of crude oil trunk lines, and one-third of refined products trunk lines (U.S. Department of the Interior, 1974). About 1.5 million barrels of oil per day is moved by tankers and barges, mainly from Gulf Coast refineries to East Coast ports (U.S. Department of the Interior, 1976b). Trucks and tank cars are used mainly for local distribution of products.

Pipeline transport is the least expensive method, provided the pipelines have a high use factor. As domestic oil production declines, the use of gathering pipelines from fields to trunk lines will decrease. Crude oil trunk lines are projected to remain in full use, since they will be used increasingly to move imported oil from Gulf Coast ports to Midwest refining centers. Net flow from the Southwest to the West Coast may be reversed to assure delivery of Alaskan oil eastward. New lines would include gathering systems for Gulf and Atlantic offshore production and, possibly, another major west-to-east line for Alaskan oil. Minor changes will naturally occur as the nation's logistic patterns shift. None of this poses insurmountable obstacles.

Regional transport needs for oil can be deduced from Table 19, which shows substantial regional imbalances in reserves, refinery capacity, and consumption.

Some changes required by these imbalances include the addition of new oil pipeline capacity from Alaska; possible movement of up to 1 million barrels per day from the Alaskan and West Coast region to the East, by either pipeline or tanker; more imports of oil to Gulf Coast area refineries (or to East Coast refineries if capacity there is increased); possible addition of pipelines to deliver shale oil to existing refineries; and additional pipeline capacity for moving imported crude oil to Midwest refineries.

Table 19 Distribution of U.S. oil resources, refinery capacity, and product consumption, by percentage of total

Region	Reserves and resources	Refinery capacity	Oil consumption in 1975
Atlantic States	2	10	36
Midcontinent	11	29	28
Southwest and Rockies	47	45	22
West Coast	16	16	14
Alaska	24	--	--
Total	100	100	100

Secondary and Tertiary Recovery of Crude Oil

Secondary oil recovery methods such as water flooding and natural gas injection are practiced nearly universally in the United States, yielding an average (primary plus secondary) recovery of 32 percent of the oil in place. Little increase in output by extension of secondary recovery methods, therefore, is expected.

Nevertheless, once all domestic oil fields are depleted, using primary and secondary methods, about 496 billion barrels will remain underground, assuming 32-percent recovery (Moody, 1975). Tertiary recovery techniques for recovering crude oil from the 68 percent remaining in depleted oil fields have been under development for many years. Such tertiary recovery methods include surfactant flooding, carbon dioxide miscible displacement, steam flood, polymer and alkaline water flood, and in-situ combustion. The Department of Energy is planning additional research in tertiary recovery. Unfortunately, these procedures are expensive, and the oil produced cannot compete at current domestic prices. If oil recovered by tertiary means could be sold on the competitive world market, many of the new methods would be economically feasible and additional production could be undertaken. Since oil recovered by tertiary means often differs significantly from that recovered by primary and secondary means, such production on a large scale would require large investments in refineries and distribution systems.

Oil Exploration as a Research and Development Priority

The development of new methods of locating oil deposits--such as the high-amplitude-signal (or "bright spot") seismic technique--should be supported through cooperative efforts by the industry and government.

As for basic geology, more information on the effects of plate tectonics and geothermal gradients on oil formation and isolation is needed. Satellite surface scanning and surveying using spectral bands, radar, and color photography are valuable in geological mapping and should be expanded; in particular, the Geosat satellite program--to cost $20 million to $50 million per year for three to five years-- should be expedited.

Work in seismic applications should include improvements in velocity determinations and more sophisticated stratigraphic analyses. In well logging, improvements in multiple, especially digital, logs can be expected.

Geochemical needs include studies of the influence of various elements and compounds, especially organics, in finding oil. Location of underwater seepages should be improved. Finally, extension of the JOIDES (Joint Oceanographic Institution for Deep Earth Sampling) program and techniques on deep ocean coring, especially on the continental margins, might be helpful.

NATURAL GAS

Natural gas accounts for about 25 percent of all primary energy used in the United States. Natural gas imports (as liquefied natural gas) are negligible at this time, but such imports are expected to increase significantly in the future. The United States produces more than 40 percent of the world's natural gas and about 60 percent of its natural gas liquids. Much of the foreign natural gas resulting from petroleum production is wasted by flaring, especially in the Persian Gulf, which is far from large markets.

Table 20 lists the estimated natural gas reserves and recoverable potential resources of the United States and the world. Table 21 represents the Oil and Gas Resource Group's estimates of how domestic reserves and resources would be affected by the conditions implied in the three scenarios for oil and gas production; these conditions are spelled out in the earlier discussion of petroleum.

U.S. Domestic Production

Domestic natural gas production peaked in 1973--three years later than that of petroleum--at 21.7 trillion cubic feet (22.1 quads) per year. By 1975, production had declined to 20.1 trillion cubic feet; the decline is expected to continue at a rate determined largely by government policies.

Again, the business-as-usual scenario assumes that there are no consistent policies for the production of natural gas. For the enhanced

Table 20 U.S. and world recoverable reserves and estimated recoverable potential resources of natural gas

Region	Recoverable reserves	Recoverable potential resources	Total
	Trillion cubic feet		
United States	216	485	701
Other market economies	1200	2692	3892
Centrally planned economies	824	2000	2824
World total	2240	5177	7417
	Quads (10^{15} Btu)		
United States	233	495	716
Other market economies	1225	2750	3975
Centrally planned economies	841	2140	2881
World total	2299	5385	7572

Table 21 Natural gas resources and reserves as affected by scenario conditions, in quads

Scenario	Recoverable reserves	Recoverable potential resources	Total recoverable resources
Business as usual	221	495	716
Enhanced supply	244	543	787
National commitment	267	592	859

supply scenarios, it is further assumed that transportation of natural gas from the North Slope of Alaska is available before 1985, and for national commitment conditions, that more gas will be recovered from tight formations.

The yearly production levels under the three sets of assumptions are listed in Table 22.

Table 22 U.S. natural gas production to 2010 by scenarios, in quads per year

Scenario	1975	1985	1990	2000	2010
Business as usual	19.7	13.5	10.3	7.0	5.0
Enhanced supply	19.7	16.1	15.8	15.0	14.0
National commitment	19.7	18.5	18.0	17.0	16.0

Imports of Natural Gas

Domestic natural gas production is supplemented by imports by pipelines from Canada, which have averaged approximately one trillion cubic feet per year from 1971 through 1977 and will likely be increased by new imports from new fields from Mexico and Canada. Imports of liquefied natural gas by ocean transport have been negligible; small but growing amounts are imported from Algeria.

Most natural gas coproduced with petroleum in other countries is flared off. In the future, however, more will be liquefied for export or made available for local use. Transporting liquefied natural gas and liquefied petroleum gas is more difficult and expensive than shipping petroleum. The amounts of liquefied natural gas and liquefied petroleum gas (in quads) likely to be imported into the United States, allowing for competition with European and Japanese markets, are illustrated in Table 23. The figures do not change after 1995 because most foreign producers will probably require increasing amounts for domestic use. These estimates do not include natural gas imported by pipeline from Canada or Mexico. Imports of Canadian gas are expected to decline; imports from Mexico may be sufficient to balance the loss of Canadian gas but cannot be estimated now.

Adequacy of the Existing Gas Delivery System

Natural gas is transported in the United States almost exclusively by 977,000 miles by pipeline. Gas produced in the Southwest is piped to all parts of the country. As domestic oil production declines, use of gathering pipelines from oil and gas fields to trunk lines will decrease. The greatest short-range needs are for a gas pipeline from the Alaskan North Slope fields to the lower 48 states and additional gathering lines for Gulf Coast offshore production. Regional imbalances between those reserves and markets that necessitate an expanded pipeline system are shown in Table 24. In addition, greater gas storage capacity is needed in the high-consumption areas of the Northeast, the Midwest, and the Great Lakes.

Table 23 Liquefied gas imports, in quads per year

	1975	1985	1990	1995	2000	2010
Liquefied natural gas imports						
Business as usual	0.015	1.7	1.9	2.1	2.1	2.1
Enhanced supply	0.015	1.9	3.2	4.2	4.2	4.2
National commitment	0.015	2.1	4.5	6.3	6.3	6.3
Liquefied petroleum gas imports						
All scenarios	0.2	0.6	0.8	0.9	1.0	1.0

Table 24 Distribution of U.S. natural gas resources and consumption, by percentage of total

Region	Reserves and resources	Consumption
Atlantic States	1	15
Midcontinent	17	32
Southwest and Rockies	66	42
West Coast	6	11
Alaska	10	--

Unevaluated and Unconventional Sources

Deposits of natural gas are found in coal mines, in geopressured hot brine deposits along the Gulf Coast, and in tight Devonian shale formations in the central United States. Although technology for extracting gas from these sources is being developed, it is not yet possible to identify potential technical and environmental problems and to estimate production levels and costs.

Of these unconventional sources the greatest deposits are in the Gulf Coast geopressured zone, where as much as 100,000 trillion cubic feet (100,000 quads) is trapped in hot brine at depths of 10,000 to more than 50,000 feet. However, only a small amount is believed to be recoverable because of geological, technical, and economic problems. Most of the gas is dissolved in brackish water or brine that would have to be brought to the surface before it could be released. The quantity of dissolved gas varies from 30 to 70 cubic feet (at standard temperature and pressure) per barrel of brine. In addition to the problem of brine disposal, drawbacks include excessive drilling depths, low well productivity, and the hazard of land subsidence in the flat coastal country already near sea level. Technology for recovering gas from these sources is not advanced, and chances of producing this gas at a cost competitive with other energy sources ($2 per million Btu) are slight.

Some gas is being produced from the extensive tight-formation Devonian shales in the central states. Again, production from single wells is low because of the low permeability of the shale. Significantly increased production would cost $2 per million Btu or more in 1975 constant dollars.

Degassing of coal seams--primarily as a safety measure for subsequent coal mining--has a well-developed technologic base, but it is little practiced because of a short supply life. Costs might be low, since they could be charged to mine safety. The primary use of shale and coal mine gas resources would be local; it is estimated that they could not supply more than 3 to 5 percent of national gas needs between now and 2010 under any scenario assumptions.

Table 25 shows a steady decline in gas availability for business-as-usual conditions, to 70 percent of the 1975 level by the year 2010. For enhanced supply assumptions, it appears possible almost to maintain 1975 production levels, and for the national commitment scenario, natural gas production could conceivably increase by 15-20 percent by the 1990's before starting to decline.

Implicit in Table 25 is the assumption that the same level of effort is applied to both increasing domestic production and increasing imports. In fact, conditions may dictate a different assumption. For example, a full commitment may be directed at increasing domestic production to minimize the need for substantial increases in imports. Conversely, a full commitment might be applied to increasing imports if domestic production falls short of expectations.

It is clear that under business-as-usual conditions, the availability of natural gas will not come close to meeting today's demands for that product. In the enhanced conditions scenario, it may be possible more or less to match current availability, but with little growth potential.

Table 25 Availability of natural gas to the United States from all sources, in quads per year[a]

Scenario	1975	1985	1990	2000	2010
Business as usual	20.9	16.8	14.0	11.1	9.1
Enhanced supply	20.9	19.6	20.8	21.2	20.2
National commitment	20.9	22.2	24.3	25.3	24.3

[a]Assumes total of Canadian and Mexican imports at one quad per year.

Table 26 adds the range of expected synthetic gas production to the estimates for natural gas availability. Again, under business-as-usual conditions, gas availability declines steadily with time, whereas under the enhanced supply scenario the total increases slightly; the national commitment conditions, however, allow for some growth.

If imports are not available in significant quantities—particularly after the 1990's—with either an enhanced supply or national commitment scenario, the most that can be produced is little more than that being produced today.

SYNTHETIC FUELS

Energy production, conversion, and consumption in the United States are influenced directly by government policies, and for synthetic fuels those policies are in a formative state. For the scenarios that follow it is assumed that foreign oil remains in adequate supply at costs between $12 and $15 per barrel (1976) until the world reserve-to-demand ratio shows signs of declining (probably about 1990-1995). In addition all scenarios are based on the assumption that the federal government will continue to support private industry development of new gasification and liquefaction process alternatives.

For the business-as-usual scenario (Table 27), it was assumed that current practices are to be continued, including regulation of natural gas and oil prices. However, federal nonrecourse, low-interest loans on synthetic fuel plants were assumed to be available for a limited number of plants (without product price support but with required

Table 26 Availability of natural and synthetic gas to the United
         States, in quads per year

Scenario	1975	1985	1990	2000	2010
Business as usual					
U.S. natural gas	19.7	13.5	10.3	7.0	5.0
Gas imports	1.2	3.3	3.7	4.1	4.1
Synthetic gases	--	0.3	1.3	3.5	4.1
Total	20.9	17.1	15.3	14.6	13.2
Enhanced supply					
U.S. natural gas	19.7	16.1	15.8	15.0	14.0
Gas imports	1.2	3.5	5.0	6.2	6.2
Synthetic gases	--	0.5	1.7	3.5	4.8
Total	20.9	20.1	22.5	24.7	25.0
National commitment					
U.S. natural gas	19.7	18.5	18.0	17.0	16.0
Gas imports	1.2	3.7	6.3	8.3	8.3
Synthetic gases	--	0.7	1.7	4.5	7.9
Total	20.9	22.9	26.0	29.8	32.2

minimum equity). In this scenario, it is assumed that the design and construction of commercial plants takes six years with initial production in the seventh. The effect of this length of construction is to reduce the rate of production of additional synthetics between 1985 and 2000.

The principal first application of coal conversion is likely to be to generate intrastate gas (onsite medium-Btu gas in the East and sale of blocks of gas under state public utility commission approval for "roll-in" to existing natural gas supplies in the West). The use of coal-derived methanol in gas turbines for peak load electricity generation is also expected. After 1995 it is likely that total business-as-usual coal production for direct combustion would be in competition with synthetic fuel production and would thereby limit production of synthetic fuels.

Table 27 Potential for coal-based synthetic fuels--business-as-usual scenario, in quads per year

Energy source	1985	1990	2000	2010
Synthetic gases	0.3	1.3	3.5	4.1
Synthetic liquids	0.1	0.3	2.3	6.1
Coal required	0.6	2.6	8.9	15.3

For the enhanced supply scenario (Table 28), it is assumed that federal deregulation of "new" oil and "new" natural gas is adopted, that the same nonrecourse loans are available as in the business-as-usual scenario, and that the existing system for permits for mining, plant construction, and operation is streamlined so that the permits are acted on within 12 months. Under these policies it is assumed that design and construction of commercial plants requires 5 years from the date of capital commitment. Assuming that the price of imported crude remains at $12-$15 (1976) per barrel through 1990, commercial plants would be used initially to augment synthetic natural gas supplies in the West and to provide modest amounts of medium-Btu gas from on-site plants in the East. Again, methanol could be used for peak load electricity generation.

Table 28 Potential for coal-based synthetic fuels--enhanced supply scenario, in quads per year

Energy source	1985	1990	2000	2010
Synthetic gas	0.5	1.7	3.5	4.8
Synthetic liquids	0.1	0.4	2.4	8.0
Coal required	0.9	3.3	9.0	19.0

For the national commitment scenario (Table 29), it is assumed that a government agency is established with authority to underwrite product prices, expedite construction and operating permits, allocate men, capital and materials, control fuel end use, and establish research priorities. Under such an arrangement, the major restraint would be the rate at which men, materials, and capital could be diverted to create a new synthetics industry. It is assumed here that $2 billion (1976) (about 6 percent of total U.S. energy-related construction) will be available in 1978 and 1979 each for synthetic plant construction, increasing at 15 percent per year until 1986, and that for subsequent years there will be an additional one-half billion per year available. It is also assumed that commercial plants can be built in four years, with coal production keeping pace with this demand. Because large amounts of water are essential to the various technologies for coal conversion, the siting of synthetic plants may be difficult in places. Although this report does not address how the water requirements could be met, the national commitment scenario assumes that there would be approximately one-million acre-feet per year available in the Missouri and Upper Colorado Basin. (See the report of the Risk and Impact Panel for a fuller discussion of this question.)

Table 29 Potential for coal-based synthetic fuels--national commitment scenario, in quads per year

Energy source	1985	1990	2000	2010
Synthetic gases	0.7	1.7	4.5	7.9
Synthetic liquids	0.1	0.7	4.7	12.9
Coal required	1.3	3.7	13.7	30.8

The projected uses for synthetics thus produced are for medium-Btu gas in several eastern locations (as 0.2 quads per year by 1990), methanol for electricity peaking in the West, substitute natural gas in the West, and synthetic distillates. After 1995, it is expected that most of the synthesis effort will be directed toward producing liquid fuels.

Potential of Oil Shale

Most of the exploitable U.S. oil shale is found along the sparsely populated, water-short Colorado-Wyoming-Utah border. Total resources in place containing at least 15 gallons of oil per ton of shale are estimated to be about 11,000 quads, of which about 3,700 quads are considered recoverable (Federal Energy, 1974). Production, however, may be restricted by state environmental standards and local water shortages. Only if great quantities of water can be imported to the area from adjacent basins can adequate production be assured. On the basis of existing pilot-plant technology, recovery of oil from shale would cost $18 to $23 per barrel in constant dollars, which, compared to domestic or world crude oil prices of $15 per barrel, is high. If, however, it is compared to other potential substitute sources, such as fuels from coal (initial selling price of $29 per barrel; or a range of $25-$35 per barrel, taking labor and engineering into account), the cost is relatively low (Hartley, 1978).

If business continues as usual, there will be no production of shale oil (Table 30). Under the enhanced supply and national commitment schedules, only moderate production is expected; the production limit, without importing water, would be 3 quads per year.

Table 30 Projections of shale oil production, in quads per year

Scenario	1975	1985	1990	2000	2010
Business as usual	0	0	0	0	0
Enhanced supply	0	0.2	0.7	1.0	1.5
National commitment	0	0.7	2.0	2.5	3.0

Use of Synthetic Crudes

Synthetic oil from coal conversion will become, along with shale oil, an increasingly large component of feeds to refineries built to receive a variety of petroleum crudes. Some research and development remains to be done on how to process these new feeds and what prerefining methods are best. Shale oil is paraffinic and, therefore, easier to accommodate; synthetic crude from coal produces significant by-product gas, which may be consumed in the plant or sold. The major changes will be needs for greater hydrotreating and catalytic cracking capacity, and development of new catalysts better suited to breaking down the heavy molecules in coal and shale oil.

A problem with synthetic liquids is the separation of particulate matter and ash. Work is needed on methods of solids removal such as filtration and agglomeration, solvent extraction, coking, and so forth (Beckner, 1976; Jannig and Bertrand, 1976).

More research, including market analysis, is needed as liquids from oil shale and coal begin to be available, especially since some feel that the use of synthetic fuels could lead to product imbalances. Such imbalances are best adjusted in the processing of natural petroleum, because trying to make synthetic crudes meet a fixed product slate could be inordinately expensive. It seems, therefore, that considerable effort will be needed to determine whether existing refineries or new refineries will be best suited to the continually changing mix of natural and synthetic crudes that will ultimately be capable of meeting market demands.

Environmental Considerations

Because coal conversion processes are still in a state of evolution, much remains to be learned about the effects that commercial synthesis plants will have on the environment. It is understandable that controversy has already risen over the potential hazards posed by such an industry and it will no doubt continue.

Coal conversion plants operate continuously under pressure, with controlled discharges released at selected locations. In many respects, the operation of these plants is comparable to that of oil refineries or chemical plants. Coal is more difficult to work with than petroleum, and the costs of preventing water and land pollution will be higher, although the technology for achieving environmental control already exists. The issue of high uncompetitive costs must also be faced as it is likely that the initial selling price of the liquids would be about $29 (1978) per barrel, or about twice the cost of imported crude delivered to the Gulf Coast.

Existing technical constraints include the need for further development of commercial-size reactors, lack of commercial-scale demonstrations for either above ground or in-situ processes, mine safety regarding both methane hazards and kerogen carcinogenicity, in-situ leakage problems, minimization of process water use, and disposal of low-Btu by-product gas. Environmental problems anticipated are air pollution, possible carcinogenicity of the spent-shale end products, and pollution of surface streams and aquifers.

OUTLOOK FOR LIQUID FUELS

The panel's projections for the availability of liquid fuels are listed in Table 31. If business-as-usual conditions persist, domestic supplies of liquid fuels are likely to decline steadily to about 60 percent of the current supply by 2010 with one-half of that supply derived from coal-based synthetics. With no change in policy, it is unlikely that conservation and fuel switching will be able to prevent at least some increase in demand for imported oil. Obviously, such a scenario creates enormous requirements for imported oil, or if it is unavailable, conversion to electricity or lower economic activity. In addition, a rapid decrease in imported oil availability such as might result from political upheaval could have serious impact on the economic well-being of the country and of most of the western world.

Table 31 Estimates of U.S. production of oil and synthetic liquid fuels, in quads per year

Scenario	1975	1985	1990	2000	2010
Business as usual					
Crude oil and natural gas liquids	20.0	18.0	16.0	12.0	6.0
Shale oil	--	--	--	--	--
Syncrude from coal	--	--	0.3	2.3	6.1
Total	20.0	18.0	16.3	14.3	12.1
Enhanced supply					
Crude oil and natural gas liquids	20.0	21.0	20.0	18.0	16.0
Shale oil	--	0.2	0.7	1.0	1.5
Syncrude from coal	--	--	0.4	2.4	8.0
Total	20.0	21.2	21.1	21.4	25.5
National commitment					
Crude oil and natural gas liquids	20.0	21.0	21.0	20.0	18.0
Shale oil	--	0.7	2.0	2.5	3.0
Syncrude from coal	--	0.1	0.7	4.7	12.9
Total	20.0	21.8	23.7	27.2	33.9

If the enhanced supply and national commitment scenarios are realized, total production from domestic sources can be maintained for the next 10 to 15 years. Thereafter, as syncrudes are produced in significant amounts, production can be expected to increase. The difference between the two scenarios is determined largely by when and how fast coal-based syncrudes are produced. In either case, it is presumed that effective energy policies that lead to supply enhancement will also reduce demand. Thus, even enhanced supply is likely to require continued and increasing dependence on oil imports for the entire period, but reduced considerably from what would occur under business-as-usual, whereas with national commitment, it may be possible to decrease imports rapidly after the turn of the century.

REFERENCES

Beckner, J. L. 1976. Trace Element Composition and Disposal of Gasifier Ash. Paper presented at Seventh Synthetic Pipeline Gas Symposium, Chicago, October 27-28, 1975. Adaptation in Hydrocarbon Processing 55(2):107-109.

Federal Energy Administration. 1974. Potential Future Role of Oil Shale: Prospects and Constraints. In Project Independence. Washington, D.C.: U.S. Government Printing Office.

Federal Energy Administration. 1976. Trends in Refining and Capacity Utilization: Petroleum Refineries in the United States; Foreign Refinery Exporting Centers. Washington, D.C.: Federal Energy Administration (PB-256 966).

Hartley, Fred L. 1978. Shale Oil: A Synthetic Fuel Whose Time Has Come. Paper presented at National Academy of Engineering Technical Session, Washington, D.C., November 2. Available from Corporate Communications Department, Union Oil Co. of California, Box 7600, Los Angeles, Calif. 90051.

Jannig, C. E. and R. R. Bertrand. 1977. Environmental Aspects of Coal Gasification. Chemical Engineering Progress 72(8):51-56.

Johnson, W. A., and R. E. Messick. 1977. The Supply and Availability of Imported Oil Through 2010. Report to the Supply and Delivery Panel, Committee on Nuclear and Alternative Energy Systems, National Research Council, Washington, D.C.

Levy, Walter J. 1977. U.S. Energy Policy in a World Context. Petroleum Intelligence Weekly, April 11 (Supplement).

Moody, J. D. 1975. Proceedings of the World Petroleum Congress.

Oil and Gas Journal. 1977. Annual Refining Survey. 75(13):97ff.

Rustow, Dankwart. 1977. Oil Crises of the 1980s. Foreign Affairs, April, pp. 494-516.

U.S. Department of Energy. 1979. Annual Report to Congress 1978: Vol. 2, Data. Energy Information Administration. Washington, D.C.: U.S. Department of Energy (DDE/EIA-0173/2).

U.S. Department of the Interior. 1974. Crude Oil and Refined Products Pipeline Mileage in the United States. Bureau of Mines. Washington, D.C.: U.S. Department of Interior.

U.S. Department of the Interior. 1976a. Commodity Data Summaries. Bureau of Mines. Washington, D.C.: U.S. Department of the Interior.

U.S. Department of the Interior. 1976b. Crude Petroleum, Petroleum Products and Natural Gas Liquids. In Mineral Industry Surveys. Bureau of Mines. Washington, D.C.: U.S. Department of the Interior.

U.S. Department of the Interior. 1977. Petroleum Refineries in the United States and Puerto Rico. In Mineral Industry Surveys. Bureau of Mines. Washington, D.C.: U.S. Department of the Interior.

Workshop on Alternative Energy Strategies. 1977. Energy: Global Prospects, 1985-2000. Massachusetts Institute of Technology. New York: McGraw-Hill.

4 COAL

Coal seems destined to become this country's principal energy resource through the year 2010, provided environmental problems can be resolved. For the generation of electricity, only coal and nuclear sources can be counted on for the next few years; both will be needed. (It is unlikely that any of the advanced renewable energy systems--solar, fusion, geothermal--can be developed and widely deployed before the next century.) Thus, coal--along with nuclear power--must play a major role.

Coal adapts itself to several uses; in addition to providing heat by direct combustion for industrial purposes and electricity production, it is a potential source of synthetic oil and gas. Although oil from shale produces a superior petroleum substitute, problems with its producibility will limit its near-term contribution, so that expected demand for fluid fuels will require both it and coal.

The U.S. Department of the Interior (Averitt, 1975) estimates that the United States has a coal resource base of almost 4,000 billion tons, or 32 percent of the world's coal resources. Recoverable reserves total about 280 billion tons (6,000 quads)--recoverable with existing technology and at current costs and prices. As exploration proceeds, recovery technology improves, and the costs of competing fuels increase, additional reserves will become available. Domestic coal production was 665 million tons in 1976. Thus, even at expanded production growth rates (6 percent per year under the assumptions of a national commitment), present and added reserves are sufficient for up to a century of use).

If coal use is to be increased, many policies and practices need to be changed. There is need for a clear and consistent federal coal policy. Leasing procedures for federal lands containing coal should be clarified; firm requirements for restoration of surface mined land must be developed; action on impact statements, advisory hearings, and permits must be expedited; and policies to make more capital funds available to coal mining and conversion industries should be promoted. Other problems include ensuring that there will be additional coal miners and mining engineers available when needed, that equipment-intensive methods will be further developed to minimize miner requirements, and that coal transport systems will be continually upgraded to prevent coal delivery from becoming a constraint on coal supply. Limits on emissions of sulfur and nitrogen oxides, particulates, and other pollutants must also be solidified.

Several CONAES Supply and Delivery Panel resource groups studied the problems in coal production and conversion; their findings are described in this section. Most problems related to increased coal use appear solvable, but only with clarification of a federal energy policy and greater cooperation among government, industry, and labor. In addition the panel feels that more capital must be made available to the coal development and transportation industry. This can only come about if coal is committed under firm, long-term contracts not subject to default because of changing demand patterns arising from changes in environmental regulations.

If there is no change in current policy, the panel estimates that annual coal production can expand to about 2,100 billion tons (42 quads) by 2010; with increased federal emphasis, production can reach about 2,400 billion tons (49.5 quads); and with a national commitment it seems possible to expand production to 5 billion tons (100 quads) by 2010, about 7.5 times the 1976 production rate of 665 million tons each year.

Increased coal use will depend on its price and the price of its products compared to the price of alternative fuels and energy sources. The real price of imported oil, as it becomes scarce, is expected to increase more rapidly than that of coal, and in spite of the fact that in the past our large reserves of coal have not been competitive with oil and gas, their escalating costs should now make coal more attractive.

RESOURCES AND RESERVES

The world resource of coal is estimated to be somewhat in excess of 16 trillion tons (Averitt, 1975). The Soviet Union contains 50 percent of the world's coal resource base; the United States, 32 percent; and the People's Republic of China, 8 percent. The distribution of the world's recoverable reserves, which total 686 billion tons (over 13,000 quads), is comparably apportioned.

The extent of coal resources and recoverable reserves, and the flexibility either to burn it directly or convert it into electricity or gaseous or liquid fuels, makes this energy source vital to the economy of the United States between now and the year 2010 (and beyond). Domestic resources, as estimated by the U.S. Department of the Interior (Averitt, 1975) total approximately 3,800 billion tons. Of this sum, 1,730 billion tons are identified, with about 747 billion tons of the identified resources (43 percent) being bituminous coal. Expressed another way, the total coal resource base is estimated to contain 80,000 quads. Perhaps half may ultimately be recoverable.

U.S. recoverable reserves--that is the quantity of coal that can be extracted and consumed within the constraints of existing technology and price--are 280 billion tons, or more than 6,000 quads. Table 32 shows the distribution of such reserves by type of coal and applicable extraction method, assuming 50 percent recovery for underground mining and 85 percent recovery for surface mining. The recoverable reserves summarized in the table are found throughout all geographic regions of the United States (Table 33).

Table 32 Recoverable reserves of U.S. coal by type of mining and type of coal

Types of coal	Underground mining		Surface mining		Total	
	(million tons)	(quads)	(million tons)	(quads)	(million tons)	(quads)
Anthracite	3,650	95	80	2	3,730	97
Bituminous	96,200	2,499	34,500	896	130,700	3,395
Subbituminous	50,100	1,002	57,860	1,157	107,960	2,159
Lignite	--	--	23,940	383	23,940	383
Total	149,950	3,596	130,700	2,438	280,650	6,034

Source: Averitt (1974)

At 1976 rates of coal production (665 million tons per year) there was a reserve-to-production ratio of 422. The comparable ratios for domestic oil and uranium for light water reactors are 8 years and 40 years, respectively.

Coal Production and Producibility

The existence of vast coal reserves and current demand for increased energy raise the questions of how fast the United States has produced coal in the past and how fast it can be produced in the future if such production is needed. As shown in Figure 16, coal production during the past 60 years has had a variance factor of about 2; however, peak

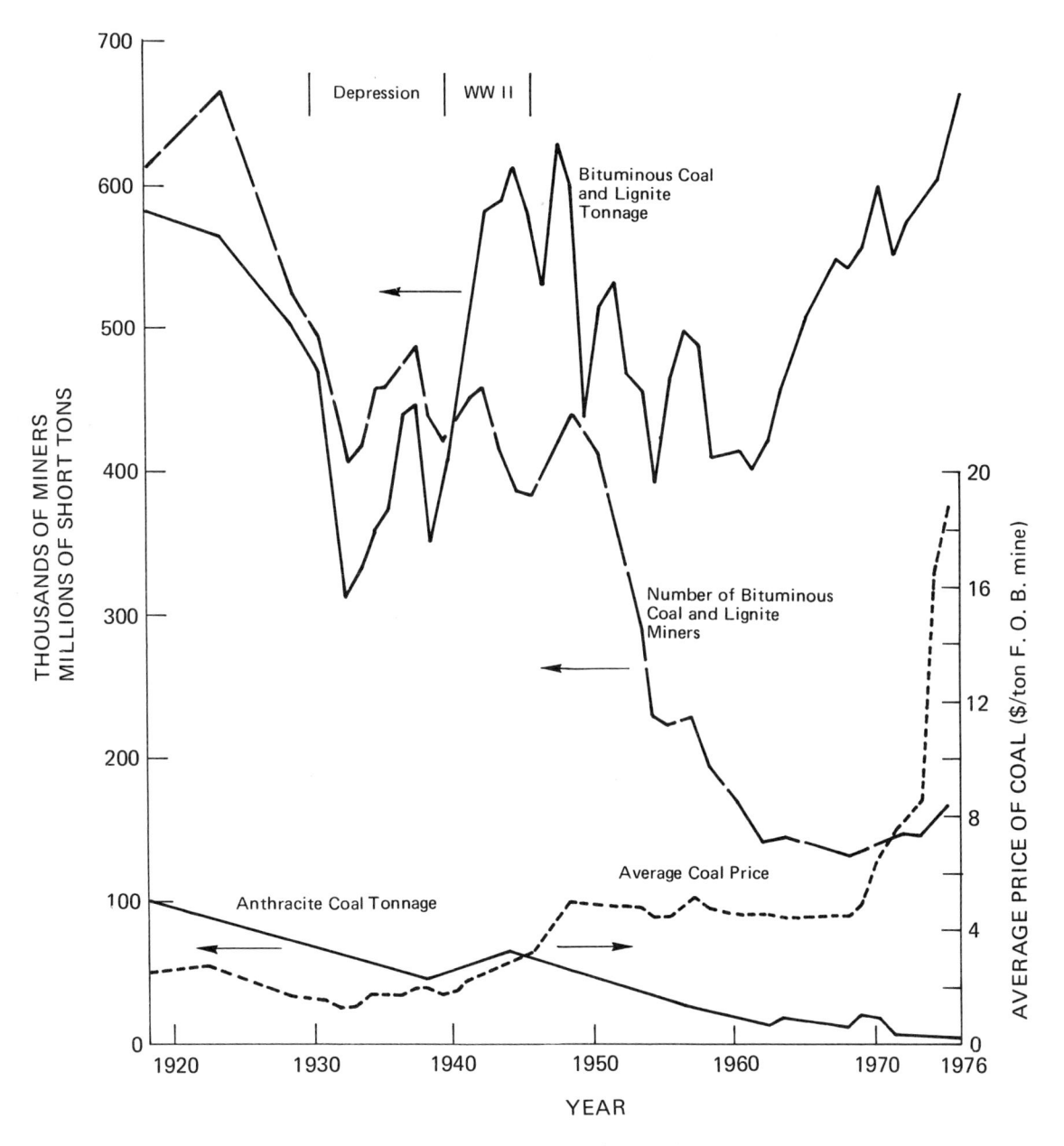

Figure 16 Domestic coal statistics, 1918-1976: production in millions of short tons, numbers of miners in thousands, and average prices in dollars per ton.

Table 33 Demonstrated coal reserve base by region of the United States January 1, 1974, in millions of tons

Mining method	East	Interior	West	Total[a]
Underground	97,456.4	81,448.9	120,934.6	299,839.9
Surface	15,826.7	26,572.7	94.486.4	136,885.8
Total[a]	113,283.1	108,021.6	215,421.0	436,725.7

[a]Data may not add due to rounding.

Source: U.S. Department of the Interior (1976)

production (including that of anthracite coal) occurred in 1918, when 680 million tons of coal were mined, compared to the 1976 figure of 665 million tons. Although the number of mines has remained fairly constant at about 8,000, the number of miners has decreased from about 800,000 in 1923 to about 170,000 today as a result of increased mechanization and automation and increased use of surface mining. Productivity increased from 4 tons per worker-day in 1918 to 20 tons per worker-day in 1969, although it has declined since to about 18 tons per worker-day. Average coal prices increased gradually from $1.31 per ton in 1932 to $4.99 per ton in 1969; then increased more rapidly to $8.12 per ton in 1973 as general inflation quickened, and finally very rapidly to $18.75 per ton in 1975 (keeping pace with the quadrupling of OPEC oil prices). Coal prices, however, vary markedly because of differences in quality, location, and transitory market conditions; mine-mouth prices currently range from $6 to $35 per ton. Transportation costs to the ultimate consumer can easily add an additional $3 to $10 per ton. The price per Btu (or per quad) has risen slightly more rapidly than the price per ton in recent years owing to greater use of western subbituminous coal and lignite. Subbituminous coal has a fuel value of 8,300 to 10,500 Btu per pound and lignite 6,300 to 8,300 Btu per pound, as compared with 10,500 to 14,000 Btu per pound for bituminous coal, which still represents the main bulk of domestic production.

Recent production gains, such as that between 1974 and 1975, have come largely from deposits in the West, which has supplied up to 42 percent of the 46-million-ton production increase. Furthermore, this increased was dominated by surface mining.

Although gains in coal production during the 1970s are significant, they do not represent full production capacity. It is estimated that 731 million tons could be produced annually at current coal prices by improved use of existing manpower and equipment. Such improvements combined with a significant real price increase could elicit a total of 782 million tons (18.6 quads) from existing mines. Coal production increases between now and the year 2010 will not depend on existing mines alone. New mines in the West, expected to be producing by 1985, can contribute an added annual production capacity of 339 million tons, while expansions of existing mines could add 133 million tons to annual production capacity. Thus, although some mines will have to be replaced, production can respond to changing supply-demand relationships (Bhutani et al., 1975).

To estimate production from now until the year 2010, three questions were examined: How would production be affected if (1) no changes were made in national energy policies and practices, and the uncertainties contained in these policies continued; (2) the federal government resolved marketplace uncertainties (for example, interpretations of environmental regulations), production uncertainties (for example, land leasing policies), and other issues dampening capital formation and, thereby coal production and use; and (3) there were a national commitment to full-scale coal energy development.

The estimates of production under sets of conditions assumed by the coal subpanel of the Supply and Delivery Panel are summarized in Table 34. With business as usual, coal production could reach 1.0 billion tons in 1985, and 1.7 billion tons by 2000. With enhanced supply efforts, production for those years could be 1.07 and 1.86 billion tons respectively, whereas with a national commitment, 1.20 and 3.75 billion tons could be produced. The lowest production estimate represents an average annual growth rate in coal production from 1976 to 2010 of 3.3 percent, the middle one 3.8 percent, and the highest of 5.9 percent.

The growth rates for coal production based on improved political and economic conditions represents a shift away from the present conditions, in which production is demand-limited to one in which production capacity would be the only limitation.

FACTORS AFFECTING FUTURE COAL PRODUCTION

To increase coal production requires raising capital to open new mines, hiring more miners and engineers, building and refurbishing transportation systems, and early solutions to many institutional and regulatory uncertainties and problems that plague the industry.

Capital Requirements

By the year 2010, under a national commitment to coal, cumulative capital needs could exceed $300 billion as old mines are replaced and new ones opened. The cumulative capital requirements shown in Table 35 are based on $50 per ton of annual capacity (1976 dollars) for under-ground mines and $35 per ton of annual capacity for surface mines, assuming

65 percent surface mining and 35 percent underground mining. It is estimated that by 2010, 43 to 51 percent of all coal production will be in the West.

Table 34 Projected annual coal production, by scenario

Year	Business as ususal		Enhanced supply		National commitment	
	(million tons)	(quads)	(million tons)	(quads)	(million tons)	(quads)
1985	1,000	19.9	1,070	21.4	1,200	24.0
1990	1,250	25.0	1,330	26.6	1,630	32.5
2000	1,700	34.0	1,860	37.2	3,750	75.0
2010	2,100	42.0	2,480	49.5	5,000	100.0

Table 35 Cumulative capital requirements for development of coal resources, in billions of 1976 dollars

Year	Business as usual	Enhanced supply	National commitment
1985	25	28	33
1990	45	49	63
2000	89	95	180
2010	139	157	305

Source: Data from Land (1975)

It should be pointed out that even in the national commitment scenario these requirements are quite modest compared with those required to utilize the coal in electric power generation or the production of synthetic fuels.

Manpower and Materials

Manpower requirements for coal production are substantial. If energy policy remains uncertain and unchanged (business as usual), there should be a gradual increase from the current level of 174,000 coal miners to 494,000 miners by 2010. Under enhanced supply development, 582,000 workers would be needed by 2010; under national commitment conditions, 1,200,000 miners would be required. Automaton and greater use of surface mining should increase miner productivity and decrease the number of miners required per ton. One additional engineer is needed for each additional 50,000 tons of coal produced annually. Thus, by 1985, 20,000 to 24,000 engineers could be required, and by 2010, 42,000 to 100,000 will be needed.

Because coal mining is a dangerous occupation, strong inducements and incentives will be necessary to attract the miners and engineers necessary to meet future needs. Over time this could result in large increases in labor costs, making coal less competitive with other sources.

By 2010 annual costs for materials are projected to be $2.4 billion, $2.8 billion, and $5.6 billion for the three scenarios.

Transportation Requirements

Transportation systems—principally the railroads but to a lesser extent inland waterways and slurry pipelines—will have to be built, improved, or expanded if coal deliveries are to increase. Requirements will depend on the combined transportation systems employed and the relative production of eastern and western coals. Studies indicate, however, that improving transportation to accommodate increased coal deliveries is feasible and practical.

Most transportation problems revolve around increasing the delivery of low-sulfur western coal to eastern markets. By 1985, about 470 new railroad locomotives and 58,000 new coal cars will be needed; an additional 13,000 rail cars will need to be replaced (Bhutani et al., 1975). More than 500 miles of new track will be required around the new mines. To keep up with increased barge traffic, most of the locks in the central river system will have to be lengthened from 600 feet to 1,200 feet, and about 243 new towboats and 1,660 new barges will be needed.

Proposals for eight coal slurry pipelines from western mines with a combined length of 4,700 miles and an annual capacity of 90 million tons are now under consideration. However, coal slurry pipelines will compete for water and other local resources with the oil shale and coal synthetics industries. Return pipelines and return water repurification may be necessary to make this mode of transportation acceptable.

Detailed projections beyond 1985 have not been made. Rather, it is essential to consider what types of planning and development can be undertaken immediately to expand the coal transportation system. Such planning must include encouragement of long-term transportation contracts and commitments, undertaking an immediate and complete inventory and review of existing distribution capability as it relates to near- and long-term energy requirements, the streamlining and integration of regulations to provide incentives for more efficient transportation, and analyses of the most efficient use of existing transportation systems for energy-related materials (including facilities and equipment).

Institutional and Regulatory Factors

As noted before, the realization of enhanced supply or national commitment coal production goals depends on changes in institutional and regulatory procedures.

The numerous overlapping and conflicting federal, state, and local regulations make quantification of the compliance costs difficult to determine. A list of such regulatory requirements for coal mines and coal-fired power plants, however, includes the permitting processes of several federal agencies and state agencies, the method of interpretation and application of environmental and worker safety regulations, and restrictions on access to federal lands. These institutional requirements cause delays. They are in fact closely interrelated because, as pointed out previously, investment in new mines and transportation equipment requires long-term contracts with utilities and other consumers. Such contracts are difficult to conclude in view of uncertainties in electricity demand projections and future environmental, health, and safety requirements for all stages of the coal fuel cycle. Ever increasing uncertainty has driven up the costs of coal production and use. These increased costs and the inconsistencies in institutional procedures make planning for new production facilities almost impossible for smaller companies and even for some large corporations.

If coal production is to be increased, the extraction of coal must increase. Research needs include the following:

- Development of less labor-intensive automated systems for extracting coal from the mine face

- Development and demonstration of systems that can move coal from the face to the preparation plant more efficiently

- Increased use of long-wall and short-wall mining systems to obtain added production and higher recovery efficiency (resource conservation) from underground coal deposits.

COAL USE

Of the 665 million tons of coal produced in the United States in 1976, 67 percent was used as fuel for electricity generation, 12 percent was used by the iron and steel industry for making coke, 9 percent was exported, and most of the rest was used by domestic industry. Only 1 percent was consumed in the commercial sector and a meager 0.002 percent for transportation. (See Chapter 2 for a discussion of the future role of coal in electricity generation.)

Industrial Use

Of the 141 million tons of coal used in 1976 by industry, 83 million tons (59 percent) were used to produce coke for the iron and steel industry. The remaining 58 million tons were used by industry primarily to produce heat and process steam. The total fossil fuels used by industry for the latter purposes was: coal, 14 percent; natural gas, 33 percent; and petroleum, 53 percent.

Use of coal to make coke for the steel industry will grow slowly in the future, for steel production is expected to grow at an annual rate of only 2 percent. Use for process heat production and steam will grow more rapidly, however, because not only will the total use of process heat increase by 2 to 3 percent per year, but coal must eventually supplant natural gas and then oil as the main industrial fuel. If coal totally replaced oil and gas in industry (excluding iron and steel use) by 2010, the growth rate would have to average 8.5 percent per year. Current consumption of 58 million tons would increase almost eighteen times to a billion tons a year for process heat and steam alone. This, of course, is based on the national commitment scenario and demonstrates how difficult it would be to rely entirely on coal. Some minor contributions from use of solar and nuclear heating for industry could ease this problem only slightly.

Feedstocks

As the supply of oil and natural gas declines, coal and shale oil will be needed as sources of raw materials for petrochemicals, industrial carbon, hydrogen, and possibly nitrogen fertilizers (ammonia, urea). Currently, about 5 percent of petroleum consumption and 3 percent of natural gas consumption are for these purposes. Before World War II, many chemicals--then called coal tar products--were derived from coal, so that a technology base exists for the transition to coal from gas and oil. This industry could revert to coal, but processes would undoubtedly be more difficult and product costs higher than at present. However, there are some new processes being developed that could improve this situation.

Exports

Much of the coal now exported from the United States is coking coal used in the steel industries of other countries, but at least a third is steam coal. In recent years, Japan and Canada have each received about a third of U.S. exports; most of the remainder has gone to western Europe and Brazil. Coal exports are expected to grow at around 2 percent annually.

Exports of coal for use as fuel or for conversion to synthetic fuels are also expected to grow as world supplies of oil and gas are depleted. Exports would be mainly to Japan and those nations developing industries that require energy resources they cannot supply alone, because western Europe probably has several decades of self-sufficiency in coal.

Coal Conversion

Little coal is currently used in coal conversion processes to produce synthetic oil and gas, because synthetic fuels cannot, as yet, compete economically with the natural fuels. As petroleum and natural gas resources become scarce, however, coal conversion is expected to grow rapidly. (See Chapter 3 for a discussion.)

These and many more research programs are essential to a long-term expanding market for coal, plus economic production of this fuel. The panel feels that it is market development, more than any other factor, that will determine the eventual contribution of coal to U.S. energy supplies.

REFERENCES

Averitt, Paul. 1975. Coal Resources of the United States, January 1, 1974. Bulletin 1412, Geological Survey, U.S. Department of the Interior. Washington, D.C.: U.S. Government Printing Office (024-001-1-02703-8).

Bhutani, J., et al. 1975. An Analysis of Constraints on Coal Production. McLean, Virginia: The Mitre Corporation (MTR-6839).

Land, George. 1975. Capital Requirements for New Mine Development. Paper presented at Third Conference on Mine Productivity, Pennsylvania State University, State College, Pennsylvania, April 1975.

U.S. Department of the Interior. 1976. Coal--Bituminous and Lignite in 1974. Washington, D.C.: U.S. Department of the Interior.

5 NUCLEAR ENERGY

INTRODUCTION

The successful introduction of nuclear power into the United States was primarily the result of the transfer and declassification of naval pressurized-water reactor technology to commercial applications; of the deployment of demonstration reactor plants on electric utility grids; and of the passage of enabling legislation such as the Atomic Energy Act of 1954 and the Price-Anderson Act.

Beginning in the early 1970's, the United States researched, developed, and implemented a strong nuclear power development program that successfully established a major new source of electric energy in a relatively short period of time. Detailed studies were performed to investigate numerous alternative reactor and fuel cycle concepts at the same time that experimental and demonstration reactors were being constructed and brought on line. Following the successful operation of several demonstration reactors in the early 1960's, Light Water Reactors (LWR's) fueled with slightly enriched uranium emerged as the primary power reactor system in the United States. Utilities made major commitments for nuclear electric generating capacity, with the annual ordering rate reaching a peak of 38 units (38,814 megawatts) in 1973. LWR's became the principal reactor type overseas as well. The United States dominated the export market with its capability to design, manufacture, construct, operate, and service nuclear power plants, as well as provide them with nuclear fuel.

This rapid evolution was made possible by general confidence in the technology's future and strong, concerted action by the government, utilities, suppliers, and engineering and construction firms. Nuclear power faced relatively little organized opposition and the promoters envisioned an unlimited future for the industry.

The 1973 peak domestic ordering rate of 38 units was followed by two sharp declines, to 17 units in 1974, and 3-5 units annually in the years 1975 through 1977. Two new nuclear plant contracts were awarded in 1978. With the cancellation of 21 reactor contracts between 1975 and 1977, the number of units on order actually declined.

Prior to 1974, United States utilities ordered more reactor capacity than all other countries combined. However, as a result of the sharp drop in domestic activity, 1974 domestic orders accounted for only 34 percent of the world total; in 1975, 28 percent; and in 1976, 12 percent. And without any extraordinary growth in foreign orders, the total foreign commitment to nuclear power (capacity in operation, under construction, or on order) at year-end 1977 exceeded that in the United States.

While much of the foreign market for reactors was originally dominated by United States reactor suppliers, foreign capability to design and manufacture reactors has increased rapidly. As utility reactor orders declined in the United States, domestic reactor suppliers lost markets abroad and United States export sales decreased from a peak of 10 units in 1974 to 2 in 1978. In addition, the recently passed legislation to control exports of nuclear materials and equipment may act to further erode the competitive position of the United States in foreign markets.

The reasons for the reluctance of United States utilities to order new nuclear capacity are nontechnical. Operating performance of LWR's has improved, and is similar to that of coal-fired units with flue gas scrubbers. Generating costs have generally remained comparable to or below those of coal-fired stations and much below those of oil-fired plants. Although nuclear power plants represent only about 10 percent of total U.S. electric generating capacity, they accounted for about 12 percent of electricity production in 1977. This reflects the utilities' preference for using nuclear plants for base-load power because of their low fuel costs.

The principal reason for the virtual moratorium on ordering nuclear power capacity can be traced directly to uncertainty about the future. This uncertainty stems from numerous concerns that, when combined, make further commitments to nuclear power unacceptably risky for utilities. Similar but somewhat less serious concerns are being expressed in the sharp decline in orders for new fossil-fired generating capacity.

A study by the U.S. Department of Energy (1978a) listed three principal concerns of utilities with regard to nuclear power:

- The perception by utilities and regulatory agencies that the Administration does not support LWR's, and thus might not carry out the federal responsibilities for licensing, spent fuel storage, waste isolation, and fuel enrichment.

- The long, uncertain schedules and escalating costs being incurred on plants now in construction or in the licensing process.

- The difficulty in securing financing and state approvals in the face of cost and schedule uncertainties stemming from the first two factors.

Specific factors contributing to the utility uncertainty described above include:

- Increasing complexity and lack of predictability in the licensing process.

- Increasingly effective opposition of anti-nuclear groups.

- Government policy changes in April 1977 deferring fuel reprocessing and plutonium and uranium recycling.

- Uncertainty of future uranium supplies, particularly in light of the deferral of fuel reprocessing and a breeder reactor demonstration project.

- Failure of government to develop and implement a plan for isolation and storage of high-level wastes.

The administration's deferral of the breeder reactor demonstration, as indicated above, is perceived by the utilities as a threat to LWR fuel supplies and the long-run viability of nuclear power. The administration rationale is that the LWR program with a once-through fuel cycle minimizes the danger of diversion and proliferation. It is the goal of the administration to stretch available uranium supplies and thereby delay the need for introduction of commercial reprocessing and breeding. In their view, LWR improvements should prove to be less risky and less expensive in the long run and would require less lead-time to introduce than would an entire new system.

Even if the utilities were willing to place new reactor orders without resolution of these issues, approvals are required from state and local utility regulators, who are often reluctant to proceed without guarantees that adequate supplies of uranium will be available and that there will be management of radioactive wastes.

This atmosphere of confusion and uncertainty threatens the survival of the U.S. nuclear industry. If remedial measures are not taken soon the United States may find itself without the option to employ nuclear energy for a substantial fraction of its additional energy needs after 1990. Since the principal obstacles blocking nuclear power are primarily institutional and largely under federal government control (licensing, reprocessing, waste disposal), the government must show evidence of a firm commitment to nuclear power before this impasse can be resolved. Until such a commitment is evident, utilities will remain reluctant or unable to order new nuclear capacity.

Recent reductions in electricity demand projections, coupled with the discovery of several large uranium deposits in Australia and Canada, have led to a speculation that advanced fission programs could possibly be delayed without undue consequences. On the other hand, there is a growing appreciation that LWR's are inefficient burners of uranium; as the world's uranium resources are highly localized, continued U.S. dependence on LWR technology could cause severe fuel supply problems for other nations. The need for a reliable supply of nuclear fuel is a strong incentive for other nations to develop fuel cycles based on reprocessing and plutonium breeders; and it is evident that a number of nations plan to do so. Fears that a commercial plutonium fuel cycle will lead to nuclear proliferation must be effectively addressed and resolved.

NUCLEAR REACTOR TECHNOLOGY

Basic Nuclear Fission Concepts

Basically a nuclear reactor is a source of high-temperature heat. In theory this heat could be used in a variety of ways. At this time, however, it is used almost exclusively to produce steam for the generation of electricity. In this application, a nuclear reactor is analogous to the boiler in a fossil-fired generating station.

The fuel in a nuclear reactor is a mixture of fissile and fertile isotopes. A fissile isotope is one whose nucleus is capable of being split by free neutrons; a fertile isotope can become fissile by absorbing a neutron. Fissile isotopes are said to be consumed or "burned" in the process of fission. When a fissile isotope splits, it generally produces two radioactive fragments or fission products, plus several additional neutrons and an enormous quantity of heat. Under proper conditions, some of the released neutrons can split other fissile isotopes, producing a chain reaction with a continuous, controlled release of energy. Both the fission products and the fertile isotopes in the fuel absorb neutrons and retard the fission process. As nuclear fuel is consumed, the quantity of fission products increases and the ability of the remaining fissile isotopes to sustain a neutron chain reaction is reduced. When this fuel "burnup" limit is reached, the reactor must be refueled.

Uranium-235 (U-235) is the only naturally occurring fissile isotope. There are two naturally occurring fertile isotopes, uranium-238 (U-238) and thorium-232 (Th-232), that can be converted to fissile isotopes. Through the absorption of neutrons, uranium-238 becomes the fissile isotope plutonium-239 (Pu-239); thorium-232 similarly becomes the fissile isotope uranium-233. To produce nuclear fuel from thorium-232, for instance, the thorium must be bombarded with neutrons from uranium-235 until enough of it has been converted to uranium-233 to sustain a chain reaction (a quantity called a "critical mass"). (Plutonium may also be used as a source of neutrons, but it too must first be created by uranium-238's absorption of neutrons from uranium-235.) Nuclear reactors fueled primarily with uranium-235 are said to operate on the uranium fuel cycle; those that use uranium-233 are said to operate on the thorium fuel cycle, in recognition of the raw material from which the uranium-233 was generated. Nearly all the reactors currently in use worldwide operate on the uranium cycle.

A fissionable nucleus is more likely to be split by a slow, or low-energy neutron than by a fast neutron. One basic reactor design, called a thermal reactor, uses a "moderator," which can consist of light (ordinary) water, heavy water (deuterium oxide), graphite, beryllium or other material with low mass number, to slow the neutrons down to "thermal" energies and thus reduce the amount of fissionable material necessary to sustain a chain reaction. Another design, called a fast reactor, does not use a moderator and employs a greater amount of fissile material than does the thermal reactor. Some fast reactors, called fast breeders, are designed to produce (from fertile material) more fissile material than they consume.

Coolant is circulated through the reactor to remove the heat produced by the fuel and to generate steam to drive the turbine generator. In certain thermal reactors such as the LWR, the light water coolant also serves as the moderator; in fast reactors, which do not use moderators, a separate fluid such as liquid sodium or helium gas may be used to generate steam.

During power production, fissile material is constantly being consumed. At the same time, fissile isotopes are being created by neutron absorption in fertile materials. In a converter reactor, such as the light water reactor, fissile material is depleted at a greater rate than it is regenerated in fertile material, making such a reactor a net consumer of fissile isotopes. Typically a converter reactor can extract only 1 to 2 percent of the energy available in the fuel, even with reprocessing. In an advanced converter reactor, fuel efficiency can be improved significantly, using 3 to 4 percent of the fuel's energy content. In a breeder reactor, fissile material is created at a rate greater than that at which it is consumed, and high-gain reactors can in principle recover about 70 percent of the fuel's energy content. An LWR with a "once through" fuel cycle (no reprocessing) uses only 0.6 percent of the energy potential of the fuel.

Spent LWR fuel contains substantial residual energy in the form of unburned fissile uranium-235 and generated plutonium-239; in reactors using the thorium fuel cycle, uranium-233 is present. These fissile isotopes can be recovered by chemical reprocessing and be refabricated into valuable reactor fuel; such reuse would substantially increase the fraction of the potentially available energy that can be extracted from the fuel, thus reducing both uranium consumption per unit of energy produced and possibly fuel costs. However, the reprocessing of nuclear fuels in the United States has been deferred indefinitely because of government concerns regarding the potential proliferation of nuclear weapons, which could result from the theft or diversion of separated plutonium from commercial fuel cycle facilities. Since all advanced converter and breeder reactors require fuel reprocessing to achieve major gains in fuel efficiency, such a policy undermines the practical feasibility of these advanced reactors.

Principal Reactor Types

In theory, a large number of distinct reactor types are possible, with many available alternatives such as the choice of coolant, moderator, fuel form, fuel cycle, physical reactor arrangement, and so on. In reality, only a few reactor concepts have been developed to the point where they can be considered as potential sources of electric energy. The following is a brief description of the principal features of the various reactor types in commercial operation today or under active development.

Light Water Reactors (LWR's)

Over 90 percent of the nuclear generating capacity in operation, under construction, or on order worldwide (over 99 percent in the United States) are LWR's, of which there are two distinct types. The Boiling Water Reactor, making up about one-third of total installed LWR capacity, is cooled and moderated by boiling light (ordinary) water and the slightly radioactive steam produced passes through to the turbine-generator. The remaining two-thirds of LWR capacity consists of Pressurized Water Reactors (PWR's), in which the coolant/moderator is pressurized to prevent boiling, and flows through a steam generator to produce nonradioactive steam.

Over the past two decades, spurred by forecasts of rapid growth in nuclear capacity with projected installed capacities as high as 1,200 gigawatts electric by the year 2000 (U.S. Atomic Energy, 1974), the United States government and industry invested more than $8 billion in developing the LWR and related facilities. The resulting combined manufacturing capacity is much larger than can be supported by the current rate of orders for reactors. This country's four manufacturers of LWR nuclear steam supply systems are capable of processing a total of about 25 to 30 units per year. The engineer-constructor firms (at least 13) and several utilities capable of designing and constructing nuclear power plants could in turn install this much capacity annually. The backlog of orders is about 143 (U.S. Department of Energy, 1978c), which provides a substantial workload to sustain the industry temporarily. However, the drop in the ordering rate over the past several years will steadily reduce this backlog to well below half the industry's capacity by the early 1980's unless the downward trend is rapidly and drastically reversed. The more pessimistic observers believe that the number of domestic orders may stabilize at five or six per year, perhaps causing one or more of the reactor manufacturers to leave the market.

Despite the worldwide acceptance of LWR's, these reactors have an inherently low conversion ratio and poor efficiency of uranium use. Uranium consumption is particularly high in this country at present due to the government-mandated use of the once-through fuel cycle (that is, with no recycling of fuel), which increases the demand for uranium oxide fuel by more than 30 percent. This abnormal mode of operation has increased the urgency of introducing more fuel-efficient reactors.

Spectral Shift Conversion Reactor (SSCR)

This is an advanced converter concept very similar to the PWR. The distinguishing feature of SSCR is the use of a varying mixture of heavy and light water as the coolant and moderator, increasing the conversion ratio and thereby reducing uranium demand over the plant life as compared with an LWR. At the same time, it complicates the reactor design and operation because of the additional equipment and operations needed for the use of heavy water and for safety. (The only demonstration of SSCR technology was in Belgium from late 1966 to 1968.) Although this concept is being restudied under the Nonproliferation Alternative Systems Assessment Program/International Nuclear Fuel Cycle Evaluation (NASAP/INFCE), there is virtually no industry or utility support for it.

Gas-Cooled Thermal Reactors

Eight different nations have shown enough interest in Gas-Cooled Thermal Reactors to construct and operate one or more units. The first to go into operation were the Magnox reactors in the United Kingdom, which were carbon dioxide-cooled, fueled by natural (metallic) uranium, and graphite-moderated. Although 26 Magnox reactors were constructed between 1956 and 1971 and operated with high load factors, their low thermal efficiency (20 percent) and low power density resulted in high generating costs. France constructed six Magnox-type reactors, which demonstrated the use of on-power refueling as well as prestressed concrete reactor vessels.

Both Great Britain and France discontinued further introduction of Magnox reactors, with France reorienting its program toward PWR's and Liquid Metal Fast Breeder Reactors (LMFBR's). In Great Britain, the Magnox reactors were superseded by the Advanced Gas-Cooled Reactor (AGR), of which two are in operation and eleven under construction. No orders, however, have been awarded for AGR's outside Great Britain, and after many years of debate the British appear to have decided to build both PWR's and AGR's in the intermediate term. As there appears to be no interest in carbon dioxide-cooled reactors outside Great Britain, this reactor type appears unlikely to play a role in meeting future nuclear demands.

High-Temperature Gas-Cooled Reactor (HTGR)

In the United States the gas-cooled reactor that predominates is the High-Temperature Gas-Cooled Reactor (HTGR), a graphite-moderated, helium-cooled advanced converter, operating on the thorium fuel cycle. A unique feature of this reactor is its use of ceramic-coated fuel, which permits the use of very high gas temperatures. These high coolant temperatures result in thermal efficiencies comparable to those of modern fossil-fired stations, sharply reducing the amount of condenser cooling water required. A long-term objective is to incorporate a direct-cycle helium turbine in the HTGR design, with a wet cooling tower that could improve plant efficiency to nearly 50 percent. The HTGR using the direct cycle would also permit the economic use of dry cooling towers, thus enabling such reactors to be sited in water-short areas. The HTGR is also a promising source of process heat at temperatures up to about $1800^\circ F$ for use in such applications as hydrogen production and direct-reduction steelmaking, provided the production facilities can be located close to the reactor. There is substantial flexibility in the HTGR fuel cycle, and the conversion ratio can be varied from about 0.6 to 0.9, depending on the fuel configuration. With the current prices of U_3O_8 and separative work, achieving conversion ratios in the upper end of this range results in increased generating costs. Although current designs specify fully-enriched (93 percent) uranium-235 (and ultimately, uranium-233), this reactor could be modified to use mediumenriched uranium fuel (only about 20 percent fissile).

Canadian Deuterium Uranium Reactor (CANDU)

The CANDU is a converter that is both moderated and cooled by heavy water, thus permitting the use of natural uranium fuel. Virtually all reactors in Canada (4,700 megawatts electric in operation or on order as of mid-1978) are of this type. CANDU reactors have also been ordered by Argentina, Korea, and Romania, in addition to the units in operation in India and Pakistan.

Coolant and fuel are housed in individual Zircalloy pressure tubes, rather than a larger reactor vessel, permitting on-power refueling on a daily basis. Operating on a once-through fuel cycle, this reactor as presently designed requires about 25 percent less uranium over its operating lifetime than an LWR operating on a once-through fuel cycle (Kasten et al., 1977).

By using slightly enriched fuel (about 1.1 percent uranium-235) rather than natural uranium, the uranium consumption of CANDU reactors could be reduced. Substantial further reductions could be achieved by using the thorium fuel cycle. Conversion ratios approaching unity appear possible, although this requires significantly reducing the core power density and thereby increasing the reactor vessel size, making the economics of this improved design uncertain. The use of enriched fuel in CANDU reactors has not been demonstrated, however, and the use of the thorium fuel cycle may require a development program of up to 20 years before the full fuel cycle can be deployed commercially. The introduction of CANDU reactors to the United States has been suggested as a possible means of uranium conservation, although considerable uncertainties remain as to whether CANDU-type reactors could be licensed in the United States in time to yield real benefits.

Light Water Breeder Reactor (LWBR)

The possibility of adapting the technology of LWR's and the thorium fuel cycle for use in a self-sustaining breeder led to the establishment in December 1965 of the Light Water Breeder Reactor Project. To confirm that thermal breeding can be achieved, the 100-megawatt-electric Shippingport PWR was modified to accommodate a new core consisting of uranium-233 and thorium. This modified reactor went critical on August 26, 1977, making it the only thermal breeder now in operation. The reactor is expected to be operated for three to four years, after which the fuel will be removed and analyzed to determine if breeding has been achieved.

Because an LWBR produces only enough uranium-233 to meet its own fuel requirements, each new LWBR requires an external source of nuclear fuel for initial core loading. This requires building "pre-breeder" reactors fueled by uranium-235 and specifically designed to produce uranium-233. Thus the introduction of LWBR's would require designing and licensing two new reactor concepts, as well as developing the thorium fuel cycle and its industrial infrastructure. Considering the lack of industry and utility interest in this concept, and considering that no formal plans to continue LWBR development beyond the Shippingport program exist, the development and deployment of LWBR's appears unlikely in the view of the panel.

Molten Salt Breeder Reactor (MSBR)

The MSBR is a proposed reactor concept that is unique in its fuel mixture of lithium, beryllium, uranium, and thorium salts. Circulated through a graphite moderator, the fuel can theoretically achieve a breeding ratio slightly greater than 1.0. This would be achieved by the use of on-line fuel reprocessing, with a sidestream of fuel being continuously withdrawn from the reactor to remove fission products.

The potential advantages of the MSBR concept are its relatively high thermal efficiency, elimination of the need for fuel fabrication, relative stability of operation, low fission fuel inventory and small waste volume. Its main disadvantages are the high corrosivity of the molten salt fuel mixture; the need for remote reactor maintenance owing to deposition of fission products on the reactor vessel, piping, and components; difficulties in containing the radioactive tritium gas produced during reactor operation; a relatively low breeding ratio as compared to fast breeders; and the need for costly materials in the reactor system. The 8-megawatt-thermal nonbreeder Molten Salt Reactor Experiment was built and operated successfully from 1965 to 1969 at Oak Ridge National Laboratory, but there is no MSBR development program at present.

Liquid Metal Fast Breeder Reactor (LMFBR)

Interest in breeder reactors in the United States began in the 1940's. The experimental LMFBR "Clementine" was operated from 1946 to 1953 at Los Alamos to demonstrate the feasibility of operating a reactor with fast neutrons, plutonium fuel, and liquid metal coolant (U.S. Department of Energy, 1978c). In 1951, the first electricity generated by a nuclear reactor was produced in the Experimental Breeder Reactor I (EBR-I) at the Idaho reactor test station (see Table 36).

Although the LMFBR can operate on either the uranium-plutonium or the thorium-uranium fuel cycle, it operates most efficiently with plutonium as the fissile material and uranium-238 as the fertile material. The reactor core holds the fuel assemblies in which mixed uranium-plutonium oxide fuel pellets are contained. The blanket region surrounding the core holds the uranium-oxide bearing assemblies.

Gas-Cooled Fast Reactor (GCFR)

The GCFR has been under development in the United States as an alternative to the LMFBR since 1961. This reactor is fundamentally different from the LMFBR in its use of helium rather than liquid sodium as coolant. Its main advantages are a potentially higher breeding ratio; a coolant that is transparent, does not become radioactive, and is chemically inert; and possible economic advantages due to potentially lower fuel costs because of superior breeding performance and lower capital costs due to simplicity of design. It also offers the long-range potential for using the helium coolant to drive a closed-cycle gas turbine, which would yield increased thermodynamic efficiencies and virtually eliminate the need for cooling water.

The GCFR is still in an early development stage and many years behind the LMFBR in technological status. Indicative of its early stage of development is the fact that no experimental gas-cooled breeder reactor has been constructed to date. On the other hand, the GCFR program is designed to take advantage of research and development in other reactor concepts, notably the LMFBR and HTGR. There is utility support for the GCFR, and it would appear prudent to continue funding its development as an alternative to the LMFBR should the latter meet unexpected technical or economic obstacles.

NUCLEAR FUEL CYCLES

Introduction

To produce the fuel needed to operate nuclear power plants, numerous processing and manufacturing steps are required. The process--known as the "front end" of the nuclear fuel cycle--begins with the extraction of naturally occurring nuclear materials from the earth's crust. Depending on the reactor design, this material must then undergo various processing stages, such as isotope enrichment, to obtain the desired fuel form. Finally, the fuel material is packaged in carefully designed fuel assemblies which must not disintegrate during extended exposure to nuclear radiation or under various mechanical stresses and chemical effects.

Following burnup of the nuclear fuel to its design limit, it is removed from the reactor core and stored on the plant site for subsequent

Table 36 World-wide fast breeder reactor plants

Name	Country	Power (MWt)	Power (MWe)	Initial operation
Decommissioned				
Clementine	U.S.	0.025	—	1946
Experimental Breeder Reactor-1	U.S.	1	0.02	1951
BR-1/BR-2	USSR	0.1	—	1956
Dounreay Faster Reactor	U.K.	60	14	1959
LAMPRE	U.S.	1	—	1961
Fermi	U.S.	200	61	1963
SEFOR	U.S.	20	—	1969
Operable				
BR-5/BR-10[a]	USSR	5/10[a]	—	1959[a]
Experimental Breeder Reactor-II	U.S.	62.5	20	1963
Rapsodie	France	20/40[b]	—	1966[b]
BOR-60	USSR	60	—	1969
BN-350[c]	USSR	1000	150	1972
Phenix	France	590	267	1973
Prototype Fast Reactor	U.K.	600	250	1974
Joyo	Japan	100[d]	—	1977
KNK-II[e]	Fed. Rep. of Germany	58	20	1977[e]

Under Construction or Hardware Committed

FBTR	India	42	17	1980
BN-600	USSR	1470	600	1980
Fast Flux Test Facility	U.S.	400	--	1979
Prova Elementi di Combustibile	Italy	140	--	1980-81
SNR-300	Fed. Rep. of Germany	760	330	1981
Clinch River Breeder Reactor	U.S.	975	350	?
Super-Phénix	France	3000	1240	1983

Planned

Monju	Japan	714	300	--
Commercial Fast Reactor	U.K.	3230	1320	--
SNR-2	Fed. Rep. of Germany	3400	1300	--
BN-1600	USSR	4000	1600	--

Footnotes for Table 36.

a Initially operated at 5 megawatts thermal as BR-5; upgraded to BR-10 (10 megawatts thermal) in 1970.

b Initially operated at 20 megawatts thermal; power increased to 40 megawatts thermal in 1970 with "Fortissimo" core.

c Also produces the equivalent of 200 megawatts electric as process steam for desalination.

d To be operated initially at 50 megawatts thermal.

e Operated 1971 through 1974 as a thermal reactor, KNK-I.

processing, or off-site storage, or disposal. This comprises the "back-end" of the nuclear fuel cycle.

This section describes the uranium-plutonium fuel cycle. Despite the fact that the thorium fuel cycle is being restudied under NASAP and INFCE INFCE, the panel considers it highly improbable because of nuclear proliferation that the United States will redirect its nuclear program toward the thorium fuel cycle. The existing nuclear fuel industry developed from a thorough comparison of alternative fuel cycles that showed that the uranium-plutonium fuel cycle is technically superior to the thorium fuel cycle. More recent analyses have, in the panel's opinion, confirmed the advantages of using LWR's with fast breeders operating on the uranium-plutonium fuel cycle. Because an acceptable solution to the nuclear proliferation problem must be found, the limited proliferation advantages of the thorium fuel cycle, if any, hardly justify establishing a new thorium-based reactor and fuel cycle industry.

The Uranium-Plutonium Fuel Cycle

The principal stages of the uranium-plutonium fuel cycle originally planned for light water reactors are presented in Figure 17, including the annual fuel material requirements for a 1,000-megawatt-electric plant operating at a 70 percent capacity factor. A description of the principal steps in the LWR fuel cycle is presented, including the current status of development and deployment.

Uranium Resources

Until breeder reactors can make a major contribution, the growth of nuclear power worldwide will be dependent on the availability of economically recoverable uranium resources and the rate at which these resources can be produced. In the event the thorium cycle is developed and becomes a major reactor fuel, thorium resource producibility will become a concern as well. In order to quantify these reserves and potential resources and assess their producibility, the Supply and Delivery Panel established the Uranium Resource Group (National Research Council, 1978). Its principal findings are summarized in this section.

According to the Uranium Resource Group, the U.S. Department of Energy provides the best systematic estimates of uranium reserves and resources. The official estimates of United States uranium-ore resources are shown in Table 37. The cost-categories of reserves are identified in terms of the highest cost in each category to produce uranium-oxide concentrate (U_3O_8) from these previously identified reserves of ore. These so-called "forward costs" are estimates of capital and operating costs not yet incurred that will be required to produce a pound of U_3O_8 at the time the estimate is made. They do not include the cost of exploration, taxes, and profits on presently identified reserves. Forward costs of concentrate per pound (in 1975 dollars) rather than market prices are used in this report except where

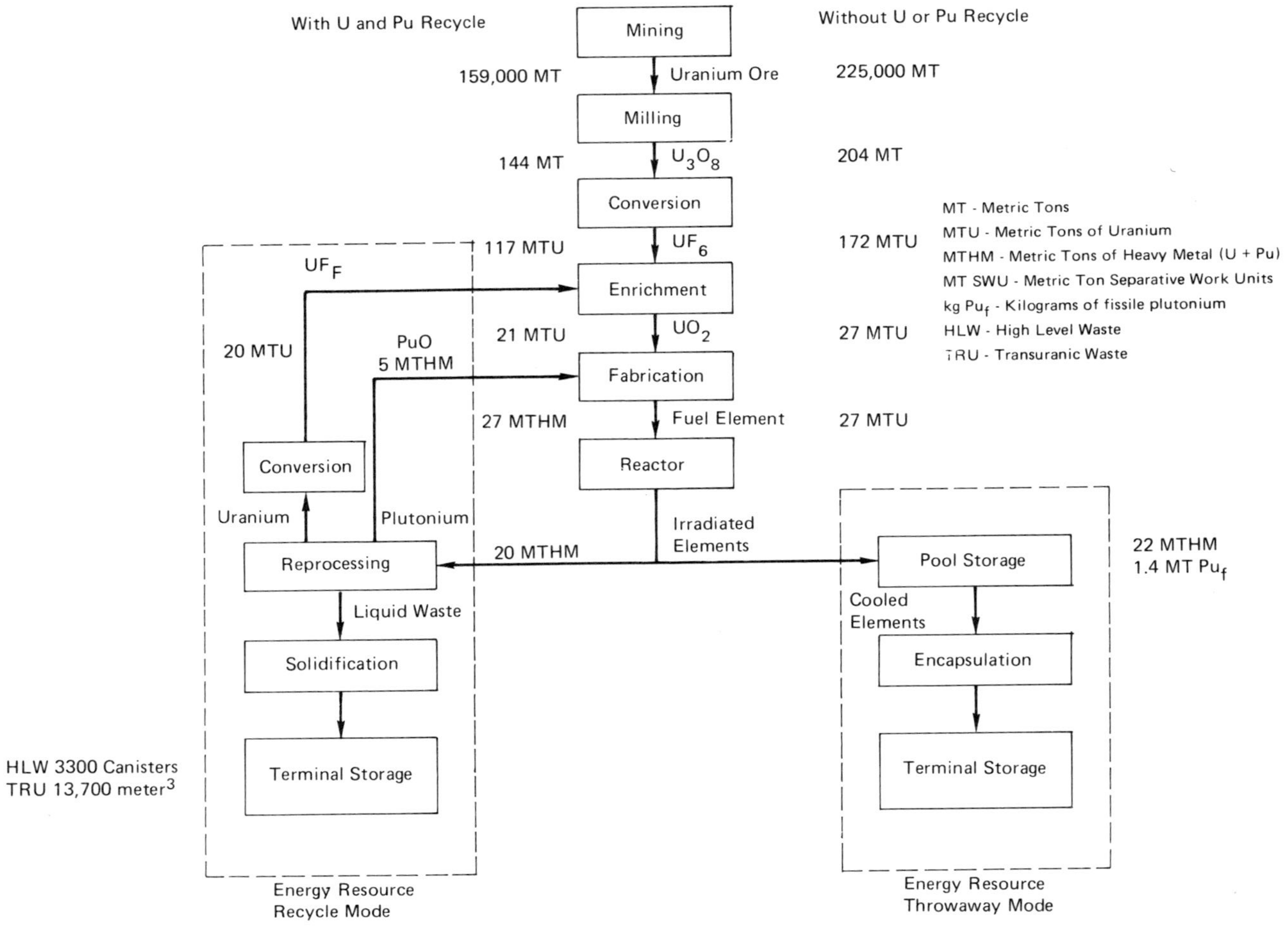

119

	With U and Pu Recycle	Without U or Pu Recycle
Uranium Ore	159,000 MT	225,000 MT
U_3O_8	144 MT	204 MT
UF_6	117 MTU	172 MTU
UO_2	21 MTU	27 MTU
Fuel Element	27 MTHM	27 MTU

MT - Metric Tons
MTU - Metric Tons of Uranium
MTHM - Metric Tons of Heavy Metal (U + Pu)
MT SWU - Metric Ton Separative Work Units
kg Pu_f - Kilograms of fissile plutonium
HLW - High Level Waste
TRU - Transuranic Waste

Energy Resource Recycle Mode

Flow: Mining → Milling → Conversion → Enrichment → Fabrication → Reactor → Irradiated Elements

From Reprocessing: UF_6 20 MTU (Uranium) back to Enrichment; PuO_2 5 MTHM (Plutonium) to Fabrication; 20 MTHM to Reprocessing.

Reprocessing → Liquid Waste → Solidification → Terminal Storage
Conversion ← Uranium ← Reprocessing

HLW 3300 Canisters
TRU 13,700 meter³

Energy Resource Throwaway Mode

22 MTHM
1.4 MT Pu_f

Pool Storage → Cooled Elements → Encapsulation → Terminal Storage

Note: The numbers in this figure represent a "snapshot" of about the year 2000. The flows of materials indicated are not intended to constitute a material balance. The numbers reflect the production quantities projected for the year 2000 and varying lead times in a growing industry.

Figure 17 Light water reactor fuel flow illustrating the average annual fuel cycle requirements per 1000-megawatt LWR for a growing LWR industry, with and without uranium and plutonium recycling.

otherwise specified. For a forward cost of $30 per pound, an approximate market price might be $40-$60 per pound.

Table 37 U.S. uranium resources as of January 1, 1976[a], in tons of U_3O_8

Forward U_3O_8 cost in dollars per pound	Reserves	Potential reserves		
		Probable	Possible	Speculative
$10	270,000	440,000	420,000	145,000
$10-$15 increment	160,000	215,000	255,000	145,000
Up to $15	430,000	655,000	675,000	290,000
$15-$30 increment	210,000	405,000	595,000	300,000
Up to $30	640,000	1,060,000	1,270,000	590,000
By-product 1976-2000[b]	140,000	--	--	--
Total	780,000	1,060,000	1,270,000	590,000

[a]U.S. resources as of January 1, 1977 (released by ERDA in June, 1977), January 1, 1978 (released May, 1978), and January 1, 1979 (released April, 1979), show changes well within the limits of uncertainty of the figures quoted here, although the category of $30-50 forward cost has been added.

[b]Estimated by-product of phosphate and copper production.

Source: U.S. Energy Research (1976)

In 1978, the U.S. Department of Energy added reserves and resources in the $30- to $50-per-pound forward cost range. The Uranium Resource Group was asked to provide resource estimates at a cutoff cost of $100 per pound, but could find no sound basis for providing a quantitative assessment at forward costs of more than $30 per pound. The U.S. Department of Energy, however, estimates that resources to $50 per pound total about 4.3 million tons.

Uranium reserves and potential resources are estimates of the quantity of economically recoverable uranium in ore that satisfies certain criteria, including minimum grade and thickness of deposits. Reserves are the most reliable class of resources with respect to location, size, grade, and economic availability, since they are based on direct measurements from drilling, radioactive logging and assaying, and sampling of the deposits. Potential (largely undiscovered) resources are divided into the three classes as defined below:

- "Probable" potential resources are those estimated to occur in known productive uranium districts in extensions of known deposits or in undiscovered deposits within known geologic trends or areas of mineralization.

- "Possible" potential resources are those estimated to occur in undiscovered or partly defined deposits in formations or geologic settings productive elsewhere within the same geologic province.

- "Speculative" potential resources are those estimated to occur in undiscovered or partly defined deposits in formations or geologic settings not previously productive within a productive geologic province, or within a geologic province not previously productive.

It was the opinion of the Uranium Resource Group that there is not, as yet, enough philosophical or geological analysis to provide a quantitative value for the total "Potential Resources." The "Possible" and "Speculative" classes, by definition, have no history of production and little identification of reserves associated with them. Hence, there is very little basis in experience for quantification.

The panel's best estimates of various classes of resources as developed by the Uranium Resource Group are given in Table 38, and are organized to distinguish reserves from potential by-product production. To assist in expressing the subpanel's uncertainty of the best value for each class, estimates of a lower limit and upper limit were derived from a subjective analysis of the available information.

Table 38 CONAES estimates for domestic uranium resources (January 1, 1976), in tons of U_3O_8 at \$30 per pound or less

Case	Reserves	By-products	Potential resources	Total
Best estimate	640,000	60,000	1,060,000	1,760,000
Lower-limit estimate	480,000	20,000	500,000	1,000,000
Higher-limit estimate	640,000	140,000	3,000,000	3,780,000

The panel's best estimate of domestic $30-per-pound uranium resources is 1.76 million tons. The lower-limit estimate is 1.0 million tons; the upper-limit estimate is 3.78 million tons. These estimates will change with time only as additional exploration and development generate new information.

Estimates of uranium resources, no matter how reliable, are not meaningful without companion information on how rapidly reserves can be identified, measured, developed, and produced. Although many studies have assumed that if a market for uranium exists, exploration will automatically respond to make the resources available when needed, this simply is not true.

The growth of uranium production depends first on the rate of discovery of new reserves. The discovery rate depends, in turn, on the kinds of incentive given industry to form and use capital for exploration, and on the state of exploration technology. The current decline in ore quality, although gradual, is expected to continue, so that the resource will become exponentially more difficult and costly to discover and produce. The Uranium Resource Group estimated the possible uranium production rates as functions of the level of commitment to discovering, mining, and processing the ore. The results of this investigation (as modified to reflect realistic nuclear industry growth rates) are shown in Table 39. Achieving such increased production rates would be formidable, but could be accomplished if government policies were adopted in support of nuclear power.

The three scenarios do not represent predictions. Rather, they are plausible levels of discovery and production under various degrees of stimulation. In fact, any number of intermediate scenarios are possible.

Uranium Mining and Milling

Uranium is found as a naturally occurring mineral in the earth's crust. Currently, economical ores (assays in the range of 2 to 5 pounds of uranium concentrate per ton of ore) are located both in deep deposits requiring underground mining and in surface deposits where conventional open-pit techniques are employed. Once mined, the ore is mechanically and chemically processed in a mill to separate a uranium salt known as yellowcake, with the chemical form U_3O_8, from the host rock and other minerals. The availability of yellowcake for the LWR fuel cycle is obviously dependent on an adequate mining and milling industry capacity as well as ample uranium reserves.

The technology of uranium mining and milling is fully developed and the existing commercial industry has a capacity of about 18,000 tons of U_3O_8 per year. This capacity is adequate to meet current requirements but a shortage of yellowcake could occur in the mid-1980's if new mining and milling facilities are not added. A lead time of approximately eight years is required to develop a new mining and milling operation and the capital investment is large. A complex with an annual production of 1,000 short tons of U_3O_8 costs about $50 million in 1975 dollars, including exploration and development.

Table 39 Potential U_3O_8 production rate, in thousands of short tons per year

Scenario	1975	1985	1990	2000	2010
Business as usual	12	24	27	33	41[a]
Enhanced supply	12	26	34	49	65[b]
National commitment	12	28	40	64	95[c]

[a] Cumulative production to 2010 is 1.0 million tons.
[b] Cumulative production to 2010 is 1.4 million tons.
[c] Cumulative production to 2010 is 1.7 million tons.

Uranium Conversion

After mining and milling, the next step in the LWR fuel cycle is the conversion of U_3O_8 into uranium hexafluoride gas (UF_6)--the form required for isotopic enrichment of uranium by the gaseous diffusion process. The technology of uranium conversion is also fully developed; the current capacity of the domestic uranium industry is approximately 22,000 metric tons of uranium per year. Such capacity is adequate to meet domestic needs over the mid-term, but proposed government plans to provide enriched uranium fuel to foreign countries could quickly alter its availability.

Uranium Enrichment

Following the chemical conversion of U_3O_8 to UF_6, the next major step is isotopic enrichment, where uranium is enriched to approximately 3 percent uranium-235. Currently, the U.S. government operates three gaseous diffusion enrichment plants whose total annual capacity is about 17 million separative work units (SWU's, a measure of the ability of an enrichment plant to perform a specified amount of isotopic enrichment). The government is now upgrading these plants to provide a capacity of 27 million SWU's by 1981. The U.S. Department of Energy has recently announced its intention to further increase U.S. enrichment capacity using centrifuge technology, which involves substantially lower electrical power consumption; it is currently planned to have a 2.2-million-SWU facility on line by 1993, with additional increments to be added as demand dictates.

This schedule for installing centrifuge enrichment capacity represents a substantial cutback from the original plan of adding 8.8 million SWU's by 1988 and results from uncertainties in future demand for government enrichment services. More importantly, it is planned to implement major power cutbacks (possibly greater than 50 percent) in the existing gaseous diffusion complex to prevent the accumulation of excessive inventories of enriched uranium. Although this policy appears sound, it raises the question of whether the large block of power necessary to operate diffusion plants at full capacity (including the additional power requirements for the Cascade Uprating Program) will be available, when needed, from the power-limited TVA system. Also, curtailing increases in capacity will limit the ability of the government to accommodate unexpected increases in demand for enrichment services, such as a shortage of uranium production capacity (which could be partly relieved by reducing enrichment plant tails concentration) or large foreign purchase orders for nuclear fuel.

Thus, despite the fact that enrichment capacity is currently in oversupply, there are concerns as to the adequacy of supply in the late 1980's. Because of the long lead times required for introducing new capacity, commitments for the next increment of centrifuge enrichment capacity should be made within the next several years.

Fuel Fabrication

The technology for fabricating virgin uranium fuel is well established in the commercial sector, and no problem is envisioned in expanding the current annual capacity of 4,500 metric tons of uranium as needed. The introduction of facilities for fabricating fuels containing recycled plutonium is currently prohibited by a government ban on the reprocessing and recycling of plutonium, and therefore cannot be resolved at this time.

Spent Fuel Storage

Fuel is irradiated in the reactor for three to four years, with one-fourth to one-third of the fuel being replaced each year. Upon removal from the reactor, fuel is temporarily stored on site in water-filled pools to allow the radioactivity and associated heat to decrease. The spent fuel is then shipped to a reprocessing plant for recovery of uranium and plutonium or to an appropriate off-site storage or disposal facility. Because government policy currently prohibits reprocessing of spent fuel, maintaining adequate space at reactor sites for spent fuel storage is becoming increasingly difficult. Despite utility plans for major expansion of on-site fuel storage basins, some reactors may have to be shut down in the early 1980's if auxiliary storage is not available. Although the government has announced plans to build away-from-reactor interim storage facilities, progress to date is not encouraging.

Spent Fuel Reprocessing

The unburned uranium-235 and plutonium by-product in spent uranium fuel can be recovered by chemical reprocessing and recycled to reduce the consumption of limited uranium resources. Using the established Purex reprocessing technology, plutonium, uranium, and fission product wastes can be separated into three output streams. Normally, the uranium would be re-enriched for use as reactor fuel, while plutonium would either be recycled in the LWR as a substitute for uranium-235 or stored for future use in fast breeder reactors.

Such reprocessing is desirable but not necessary for LWR operation; it is absolutely essential for all advanced reactors using plutonium or uranium-233 as fissile material. Since advanced reactors using these man-made fuels are essential to increase the efficient use of uranium, reprocessing will be needed if nuclear power is to contribute significantly to the U.S. energy supply beyond the end of this century.

Uranium-plutonium reprocessing technology developed by the government is directly applicable to commercial reprocessing of LWR fuels. Three commercial reprocessing plants have been built in the United States. The Nuclear Fuel Services plant at West Valley, New York, was operated for a number of years, but changes required by the government during a facility expansion program compromised its economics and the plant was shut down. The General Electric plant at Morris, Illinois, was designed and built on the basis of a new and unproven reprocessing technology and experienced technical difficulties during cold testing; it appears unlikely that this small (one ton of heavy metal per day) facility will ever operate. Construction of the largest of the three (five tons per day), by Allied Gulf Nuclear Services, at Barnwell, South Carolina, is nearly complete, requiring only some additional equipment for treating radioactive gases and converting plutonium to the oxide form required for shipping and fuel fabrication. Completion of this plant has been deferred indefinitely as a result of the government-imposed moratorium on fuel reprocessing.

Outside the United States, reprocessing plants are operating in Great Britain, France, Belgium, and Japan. Although existing capacity is relatively small today, Britain, Japan, France, Germany, and Brazil all have announced plans for large commercial-size reprocessing facilities. These plants may become international reprocessing centers subject to international safeguards to minimize the risk of theft or diversion of plutonium or uranium-233 for use in weapons, although weapons programs already exist in Great Britain, France, and Brazil.

Disposal of High-Level Radioactive Wastes

A substantial amount of successful development work on the ultimate disposal of high-level radioactive waste has been completed, and the remaining effort required is mainly the application of the available information and technologies in an actual waste repositor to demonstrate that these wastes can be disposed of permanently and safely.

The design, construction, and operation of the Waste Isolation Pilot Plant (WIPP) intended for disposal of radioactive wastes from military programs is an important step in demonstrating that a solution of the waste disposal problem exists; however, it is also important to proceed with a similar demonstration of the technology required for disposal of commercial wastes, whether or not this is accomplished in the WIPP facility, as is discussed in a later section on radioactive waste management.

The Thorium Fuel Cycle

The various steps comprising the thorium fuel cycle proposed for use in advanced converters are depicted in Figure 18. Because uranium-233 does not occur in nature, advanced converters must initially be fueled with a mixture of enriched uranium-235 and thorium. The spent fuel must be reprocessed periodically to recover uranium-233, produced from the thorium. Since these reactors have conversion ratios less than 1, they would not be self-sustaining in uranium-233, and it would be necessary to make up for depletion of the fuel with uranium-235 until enough uranium-233 has accumulated to fuel the entire core.

If a thermal breeder is eventually developed, it is envisioned that separate pre-breeders optimized for high uranium-233 production would be built and operated as "fuel factories" to supply uranium-233 to the thorium breeders. Thus, two different types of fuel would have to be fabricated as well as reprocessed. Since uranium-238 will be present in the enriched uranium-235, pre-breeder fuel and subsequent breeder fuel will also be contained in the spent fuel of these reactors and require special handling and processing. Table 40 lists the various reactors which have actually operated on the thorium fuel cycle.

Thorium Resources

At this time, the uses of thorium are few; there is little demand and even less production. It is being used as an additive to metallic tungsten filaments for lighting and welding, as an additive to magnesium-based alloys to increase their strength at high temperature, and as a primary electron emission source in certain electronic devices. As a result of its limited commercial use, little exploration has been done that would provide estimates of our domestic thorium resources.

How much undiscovered economically recoverable thorium exists is virtually unknown compared to the extensive geologic data that exist for uranium. In fact, there are no thorium mines in operation today and the thorium produced to date has been obtained as a by-product of two titanium mines in Florida. The fact that the crustal abundance of thorium is approximately four times greater than that of uranium has no substantial bearing on the potential size of economically recoverable thorium resources.

Based on the limited amount of information available, the Uranium Resource Group estimates that potential domestic resources of ThO_2 are

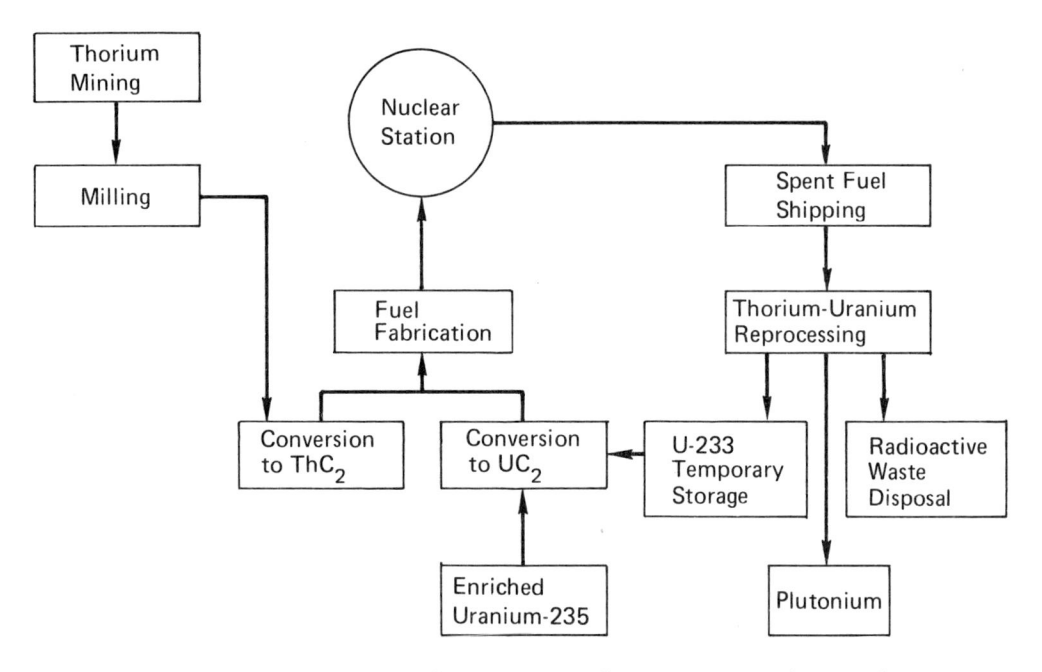

Figure 18 Proposed thorium fuel cycle for use in advanced converters.

Table 40 Experience with thorium/uranium-fueled reactors

Reactor	Location or sponsor	Experience
Advanced thermal reactor	Federal Republic of Germany	15-megawatt-electric helium-cooled "pebble bed" reactor; operated with fuel pebbles containing microspheres of thorium-uranium carbide in graphite.
BORAX-IV	Idaho National Engineering Laboratory	Thorium oxide was included in boiling water reactor core experiments.
Dragon	United Kingdom	20-megawatt-thermal helium-cooled reactor using uranium fuel and thorium fertile material; operation beginning in 1964.
Elk River	Elk River, Minn.	Boiling water reactor operated with thorium oxide in core.
Fort St. Vrain	Near Denver, Colo.	330-megawatt-electric high temperature gas reactor thorium converter; is in power ascension stage.
Indian Point No. 1	Buchanan, N.Y.	Commercial pressurized water reactor that used pellets of urania-thoria in first core; 1962.
Light water breeder reactor	Near Pittsburgh, Pa.	Seed-blanket core similar to light water breeder reactor; built and operating in Shippingport Atomic Power Station.
Molten Salt Reactor Experiment	Oak Ridge National Laboratories, Tenn.	Molten Salt Reactor Experiment was first reactor fueled exclusively with U-233; 1968.
Peach Bottom No. 1	Peach Bottom, Pa.	40-megawatt-electric gas-cooled reactor fueled by carbon-coated uranium-thorium carbide microspheres; operated 1966-1974.

about 500,000 tons in deposits of uncertain economics. Although this potential thorium resource base is substantially less than the group's "prudent planning" estimate of 1,800,000 tons of U_3O_8, thorium consumption in nuclear reactors would be relatively small. For example, in an HTGR with a conversion ratio of 0.65, total thorium requirements over the thirty-year plant lifetime of a 1,000-megawatt plant would be less than 300 tons, without recycling the thorium recovered during the reprocessing of spent fuel to recover uranium-233. In practice, recycling of recovered thorium would reduce the lifetime requirement to less than 100 tons. This is less than 2 percent of the uranium demand of an LWR operating with recycling of both uranium and plutonium. Thorium resource availability is therefore not considered to be a significant concern despite the absence of thorium resource data. Of much greater concern is the fact that the various technologies required to establish a commercial thorium fuel cycle are in a very early stage of development and that industrial and utility interest in this technology is minimal.

Fuel Cycle Status

The fact that electricity generation from nuclear power plants may be constrained by the availability of uranium resources as well as by proliferation concerns associated with plutonium fuels has led to renewed interest by the United States government in the thorium fuel cycle. Experience for the fabrication of thorium fuels is limited. The most recent application of thorium has been at the Fort St. Vrain HTGR plant and for the LWBR core for the Shippingport reactor. The Fort St. Vrain fuel is graphite-coated ceramic fuel unique to the HTGR, whereas at Shippingport the fuel is metal-clad oxide. As yet, there is no precedent for fabricating recycled uranium-233 that has been irradiated to high burnup levels. This fuel contains significant quantities of radioactive uranium-232, which necessitates performing the fabrication operation remotely, in a shielded enclosure.

The reprocessing of thorium fuels containing uranium-233 has not yet been demonstrated. Thorium fuels are generally somewhat more difficult to dissolve, and if thorium and uranium are reprocessed separately, as in the HTGR, additional reprocessing steps are required.

The essential portions of the thorium/uranium-233 cycle have not been demonstrated on an adequate basis to encourage support for the large scale industrial interest and commercialization of thorium fuel cycle facilities. At this time the government has a thorium fuel cycle development and demonstration program which, if implemented and successful, will provide fuel reprocessing and fuel refabrication data sufficient to design production scale facilities. In general, the work completed to date on the fuel cycle has involved preliminary studies and conceptual design, not pilot plants. Considering the extensive amount of effort that was required to establish this processing technology for uranium-plutonium fuels, reaching a similar stage of development for the thorium cycle may require up to 20 years of effort.

Nuclear Fuel Cycle Economics

In spite of the relatively high capital costs of nuclear power plants (compared to fossil-fired units), nuclear energy is economically attractive because nuclear fuel cycle costs are low. This is demonstrated in Table 41 which gives a breakdown of generation and fuel cycle costs of an LWR authorized in 1979 and beginning commercial operation at a midwestern location in 1990, based on current trends in financial parameters and construction schedules (Brandfon, 1978). This information reveals that the overall fuel cycle cost would represent only one-third of the total generating cost and, furthermore, that the price of yellowcake is a comparably small fraction of the total fuel cycle cost.

The sensitivity of LWR economics to uranium price is illustrated in Figure 19, where it is seen that the current spot price of yellowcake ($43 per pound of U_3O_8) yields a nuclear generation cost (86.4 mills per kilowatt-hour) substantially lower than the generation cost of a midwestern coal-fired unit (93.4 mills per kilowatt-hour) using lowsulfur coal at current coal prices ($1.30 per million Btu). A 50 percent increase in the current price of uranium (to $63 per pound of U_3O_8) would increase nuclear generating costs by only 6 percent, resulting in approximately the same generating costs as those of coalfired plants. Although the current market price of uranium provides a substantial competitive edge over other fuels, the magnitude of future uranium price increases will determine how long LWR's remain economic.

As installed nuclear capacity approaches the nominal 300 LWR plants that can be fueled by this quantity of low-cost uranium with the once-through cycle (100 plants if only proven uranium reserves are available), then exploration for and production of less reliable and more costly uranium will be required to permit continued growth in nuclear capacity. There is a body of opinion (Nuclear Energy, 1977; U.S. Department of Energy, 1978b) that such exploration can be confidently expected to yield substantial quantities of new uranium, as needed, at modest increases in forward cost.

NUCLEAR POWER ISSUES AND CONSTRAINTS

In spite of what the panel believes is the established need for nuclear power to satisfy current and future energy requirements, several issues have become obstacles to public acceptance of nuclear power. In addition, government constraints on the use of nuclear energy have inhibited electric utility commitments to nuclear plants. These issues and constraints need not undermine nuclear power as a component of our energy system if effective measures are taken to help the public become better informed on the relative merits and risks of this technologically sophisticated energy source. The federal government must also make a firm commitment to reduce unnecessary political and institutional barriers.

Unnecessary delay in commercial demonstration of a promising energy technology that has reached the advanced development stage now occupied by the LMFBR is not only a waste of valuable human and financial resources but also a senseless restraint on future energy supplies. It

Table 41 Light water reactor fuel cycle and generating cost components, in mills per kilowatt-hour[a]

Thirty-year levelized fuel cycle cost

Yellowcake at $43 per pound U_3O_8	10.6
Conversion at $2.15 per pound U	0.5
Enrichment at $100 per separative work unit	5.8
Fabrication at $125 per kilogram	1.6
Carrying charges at $440 per kilogram	6.7
Back-end charges at $125 per kilogram	2.3
Total	27.5

Thirty-year levelized generating cost

Fixed charges (investment)	53.1
Fuel	27.5
Operation and maintenance[b]	5.8
Total	86.4

[a]Assumptions:
1100 megawatt light water reactor plant--30 year service life with 65-percent average annual capacity factor
Once-through fuel cycle
Eleven-year plant design and construction period
Plant capital costs in 1978 dollars escalated at 8-percent per year for first two years, 7-percent per year for the next five years, 6-percent per year thereafter
Plant startup in 1990
Lifetime levelized fixed charge rate for plant: 15.7-percent per year
Fuel cycle and operation maintenance costs commence in 1990, 1978 dollars escalated at 6-percent

[b]Includes nuclear insurance and decommissioning

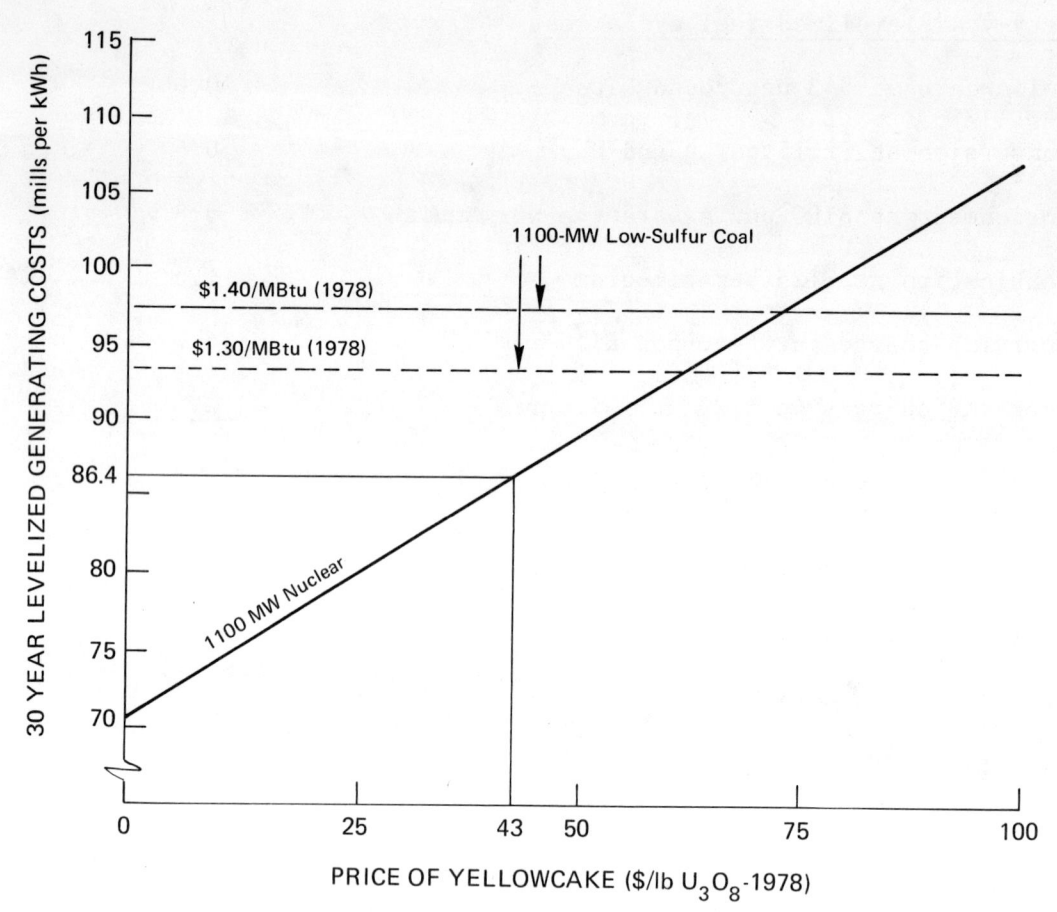

Figure 19 Sensitivity of light water reactor generating cost to the price of uranium, with generating cost estimates for Midwest coal-fired generating unit as functions of the price of low-sulfur coal (dashed lines) added for comparison.

is in the context of these and other practical considerations that near-term decisions should be made concerning development and subsequent deployment of advanced reactor systems and their fuel cycles.

Uranium Supply

The availability of a reliable supply of nuclear fuel at a reasonable price is a basic requirement for the long-term viability of nuclear power. Undue optimism with regard to the recovery of uranium ore from undiscovered deposits could lead to future fuel shortages at operating nuclear plants, whereas exclusive reliance on the relatively modest inventory of proven reserves could discourage electric utilities from adding new nuclear generating capacity. The panel believes that rational policy with regard to uranium supply would be to plan for an LWR industry based on known and probable uranium resources of 1.8 million tons of ore; that supply would be sufficient to satisfy the lifetime fuel requirements of approximately 300 LWR's using the oncethrough cycle and would justify increasing reliance on conventional nuclear plants until the end of the century. At the same time, the federal government must continue its exploration activities to reduce the insecurity of a dwindling United States uranium resource base and support necessary R&D of advanced nuclear systems that can use limited uranium supplies more efficiently.

It should be noted that inefficiently burned uranium is not lost forever as is the case with inefficiently burned oil or coal. Plutonium and uranium-235 can be recovered from spent fuel at some point in the future, and additional LWR fuel can be stripped from diffusion plant tails through laser isotope separation. There needs to be an estimate of the total fuel extension possible from these two processes, i.e., reprocessing existing spent fuel and stripping existing tails; the latter is about 60,000 tons of U_3O_8 equivalent. In the future the two techniques together could nearly double the resources available from a given amount of U_3O_8. All this is quite apart from the fuel value of uranium-238 in tails when they are burned in breeders at some future time.

Nuclear Proliferation

The possibility that unstable national or terrorist groups might acquire nuclear weapons has been a major concern since the Manhattan Project. Initially, to guard against this, tight security was imposed on both materials (nuclear fuel) and information. Today, 35 years later, the physical security program for weapons has been—as far as is publicly known—highly successful, although information concerning nuclear technology is now well-known through the world.

In the last few years, it has been noted that the material required for nuclear weapons production could, in principle, be stolen from civilian nuclear power programs, which thus might directly contribute to the proliferation of nuclear weapons. Partly in response to this potential threat, the current administration ordered an indefinite delay in fuel reprocessing and a redirection of the breeder reactor program, which has resulted in construction delays of the next prototype reactor (the Clinch River project), and instigated an International Nuclear Fuel Cycle Evaluation (INFCE) to search for a proliferation resistant fuel cycle. INFCE is to explore new institutions as well as new technologies. It is not just a search for a technical fix, but an effort to identify all factors which may influence timing and management of nuclear fuel cycle developments.

At this point, the purpose and scope of a "proliferation-resistant" fuel cycle should be considered. The essential objective is to make the diversion, or theft, of nuclear fuel by small groups (terrorists, anarchists) as difficult as possible, if not impossible. Although some observers have maintained that a proliferation-resistant fuel cycle should also decrease the probability of sovereign national states obtaining weapons fuel, this goal simply is not tenable--the technology for designing fuel production reactors and fuel reprocessing plants is known throughout the world. For example, the United States went from the demonstration of the fission chain reaction (Fermi Pile, December 2, 1942) to the first plutonium explosive (Trinity Test, July 16, 1945) in two and one-half years, beginning with virtually no nuclear technology. Other nations, beginning with a highly developed technology, should now be able to do the same. In spite of the fact that most technically sophisticated nations could produce nuclear explosive material by cheaper, faster, more direct and predictible methods, some experts have argued that the mere presence in a country of a nuclear reactor using fuel that could be employed in an atomic explosive would pose an intolerable temptation for confiscation and weapons production by the host country.

All reactor types generate approximately the same quantities of fissile plutonium (Figure 20), and therefore we should concentrate on the fuel cycle rather than the reactor for potential solutions. Weapons require the use of highly concentrated fissile material, and we should consider what intrinsic properties we can give to the fuel cycle so that fuel diversion and conversion into a concentrated fissile form becomes more difficult. Four conceptually distinct approaches, and combinations thereof, are possible:

- Colocation of fuel cycle facilities
- Chemical dilution
- Isotopic dilution
- Radioactivity spiking

The most vulnerable link for fuel diversion is between fuel reprocessing and fuel fabrication. At this stage the fissile fuel contains little radioactivity, and may exist in a relatively concentrated form. To minimize the probability of diversion in this part of the cycle, reprocessing and fabrication plants could be located in common, secured sites. Other facilities, such as plutonium storage and waste disposal, could also be part of such "fuel cycle parks."

Indeed, the concept of a secured nuclear park could be extended to include the reactors themselves. However, since one reprocessing plant could serve as many as 50 reactors, it would be desirable to have many more "reactor parks" than "fuel cycle parks." One strategy would be to confine reactors to the 100 or so existing or planned sites. These "mini-parks" would allow electric power utilities adequate flexibility in plant siting and, at the same time, permit the establishment of secure and permanent nuclear centers to minimize safeguard problems. In a study of energy centers in 1976, General Electric concluded that attributes of construction and fuel management in parks could warrant further study that would ultimately include breeder reactors (Finger, Lieberman, and McNelly, 1976).

If fissile fuel could be maintained in a dilute form the task of diversion and weapon fabrication would become somewhat more difficult since chemical separation of fissile elements would be required. For thermal reactors, which require dilute (2 to 4 percent fissile) fuel, this is easily achieved. The fuel needs for future breeder reactors, which require 10 to 25 percent fissile concentration, could also be satisfied if the fuel outputs from the reprocessing plants were restricted to 25 percent fissile (for example, 25 percent plutonium in a plutonium-uranium mix). Thus, fuel dilution is a readily constructed barrier against diversion and illicit weapons fabrication.

A better safeguard protection would be to mix the fissile isotope with nonfissile isotopes of the same element. Thus, the difficult task of isotope separation would be necessary to obtain concentrated fissile fuel. In this respect the current LWR fuel, low enriched uranium (uranium-235 in uranium-235/uranium-238), is a highly proliferation-resistant fuel, and this safeguarded fuel would provide the bulk of the world's nuclear fuel supply for the next several decades.

Only two breeding cycles are possible: the uranium-plutonium cycle, and the thorium-uranium cycle. For each of these cycles, the initial fertile isotope (uranium-238 or thorium-232) is of a different element from the final fissile isotope (plutonium-239 or uranium-233). Although neither cycle by itself can provide isotopic dilution, limited isotopic dilution safeguards could be achieved by mixing the fissile uranium-233 from the thorium-uranium cycle with the fertile uranium-238 from the uranium-plutonium cycle.

What the above implies is that a reactor development scenario could be envisioned in which some reactors use isotopic diluted fuel. Since uranium-233 mixed with uranium-238 is an attractive fuel for thermal reactors, the following symbiotic relationship can be postulated. First, uranium-233 is generated in the blanket of breeders, diluted with uranium-238, and used as a safeguarded fuel in thermal reactors. Thereafter, the plutonium generated in the thermal reactors is used to fuel breeder reactors.

It is understood that any nuclear reactor fuel can be converted into an explosive weapon by a team with technological and scientific knowledge backed by large resources. No "technical fix," or technical alteration to the nuclear fuel cycle, can absolutely prevent the possible conversion of reactor fuel into a crude nuclear explosive, and

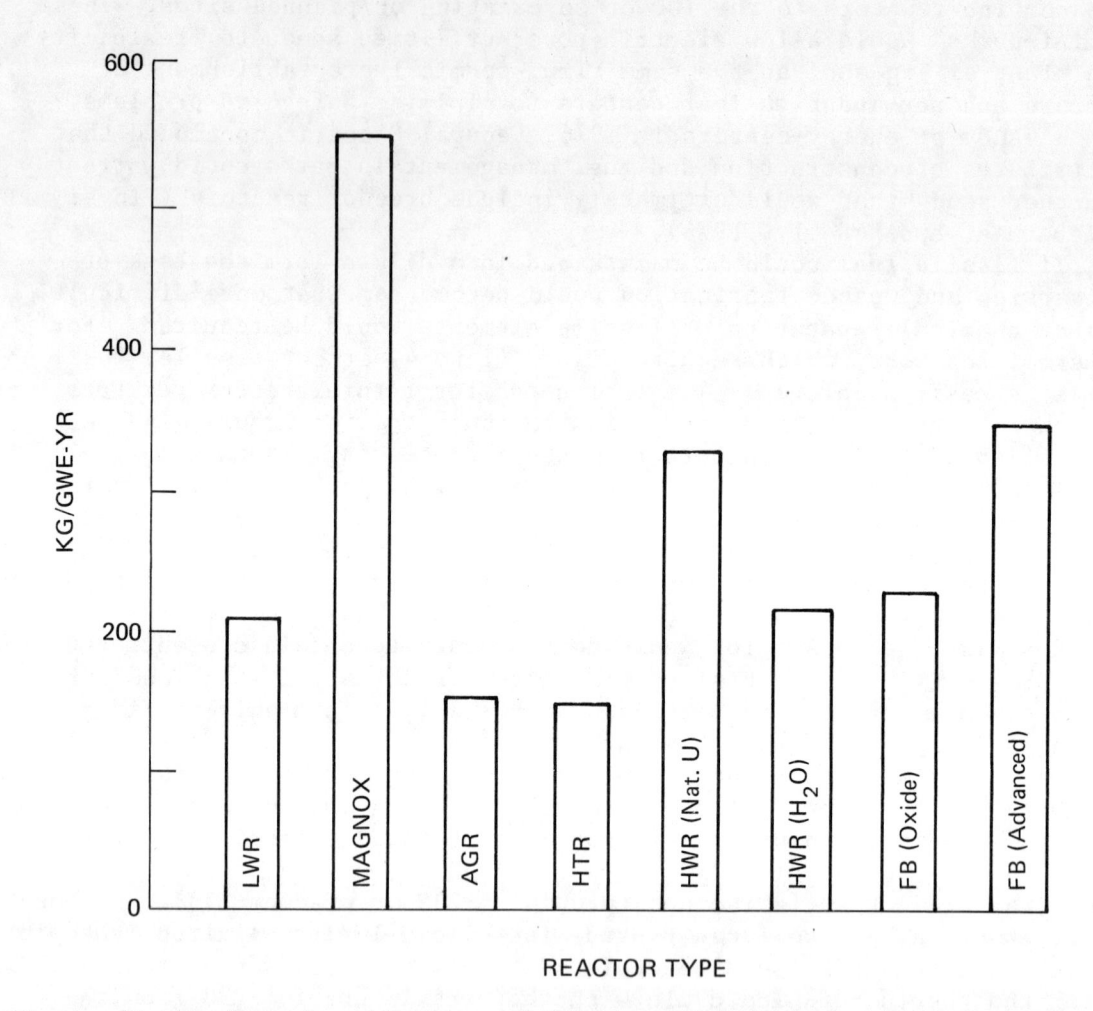

Figure 20 Estimated production of fissionable plutonium in representative power reactor types, in kilograms per gigawatt-electric per year.

the primary safeguard against diversion is, and must remain, physical security.

In an address to a joint United Kingdom-United States nuclear fuel cycle meeting in London in 1978 Sigvard Eklund, director-general of the International Atomic Energy Agency, expressed concern that, rather than serving to tighten controls, the nonproliferation policies of the nuclear exporting nations are exacerbating what has become a tense policy standoff (Energy Daily, 1978). The United States is seen as the worst offender, leading to a climate of total distrust in the international community.

It should be observed that the real risk humanity faces is not nuclear fuel proliferation at all, but rather nuclear war based upon the current awesome stockpile of nuclear weapons—a war perhaps triggered by the ever-dwindling supply of fossil fuel. Individuals, rich or poor, acting separately or as nations, will not be content to sit idly by and watch their standard of living slowly decline. Indeed, to set energy policies that offer no hope for future abundance is probably a greater threat to world stability than any posed by the widespread diffusion of nuclear technology.

Radioactive Waste Management

Radioactive waste management has become one of the prime issues in the debate on the future of nuclear power in the United States. The public is generally concerned about the availability of methods for safe disposal of commercial wastes. Although it is assumed by many in the nuclear industry that the technology for the safe disposal of high-level radioactive waste is available, a demonstration program is needed to begin to find answers to the many technological questions associated with permanent disposal of nuclear wastes. The United States must deploy a disposal system that uses current technology and provides adequate protection of public safety and health to meet initial needs while research continues to search for improved and more cost-effective methods. In the near term, adequate spent fuel storage capacity should be provided to avoid fuel management problems at plants that could lead to forced reactor shutdowns.

Prime responsibility for the disposal of radioactive waste lies with the federal government. Up to the present time, federal waste management programs have been unable to establish policies and strategies to demonstrate and deploy repositories for ultimate disposal in a technically and politically acceptable way. As a result, several states have laws restricting the establishment of disposal sites and others are considering such actions.

Most experts agree that permanent geologic disposal of wastes is an acceptable solution, although uncertainty remains as to the preferred geologic medium, such as salt or basalt. Even though additional research is desirable to determine the optimum medium for disposal, it is nevertheless important for public acceptance of nuclear power to proceed with the design and construction of a repository now, based on adequate technologies. This repository might include provisions for intact spent fuel assemblies, but its primary function should be the permanent disposal of wastes resulting from the reprocessing of spent fuels, with retrievability as a key feature of the initial design, so that any unanticipated disposal problem can be accommodated safely (Waste Solidification, 1978).

Another solution to the waste problem, involving various established technologies, is being pursued in foreign countries such as West Germany, Canada, and Sweden (Wivstad, 1978). This solution consists of confining the radioactive waste using several barriers: immobilization of the waste in a chemically inert structure such as a glass or ceramic matrix, packaging of this solid form in an engineered container, and deep burial of the package in an appropriate geologic medium to isolate the waste from the biosphere.

Nuclear Licensing

Only four nuclear plants were ordered domestically in all of 1977 with two plants being purchased in 1978. Furthermore, several nuclear plants in the construction stage, such as the Seabrook Nuclear Plant in New Hampshire, have experienced lengthy delays. This construction slowdown has generally not had an adverse impact on the current availability of generating capacity because the recent modest increase in load growth had been adequately accommodated by existing reserve margins. However, timetables for installation of new nuclear plants are being pushed farther and farther into the future, and utilities are no longer able to plan effectively.

However, the long and uncertain lead time in bringing a nuclear power plant on line is generally considered to be the major factor inhibiting electric utilities from making further commitments to nuclear power. Utilities must now allow 12 to 14 years for bringing a nuclear power plant from initial planning to commercial operation. This costly delay results from a cumbersome licensing process involving more than 50 federal and state agencies. Although the objectives of nuclear facility licensing—such as public health and safety and preservation of environmental quality—are basically sound and necessary, licensing has become an unwieldy process that threatens the viability of nuclear power itself.

Fortunately, industry and government efforts have been initiated to simplify the nuclear licensing process and to remove redundant regulatory constraints. The political, technical, and social aspects of this important task will present substantial difficulties and challenges that must be addressed directly and expeditiously if nuclear power is to have a major supply role in the future.

ADVANCED REACTOR SYSTEMS

Practical Realities of Advanced Reactor Deployment

Despite the fact that LWR's are expected to provide nearly all the nuclear power produced in the United States for the remainder of the century, their potential contribution is limited by the quantity of the uranium resource base (Table 42).

Table 42 Energy recoverable from domestic uranium resources

CONAES Uranium Resource Subpanel estimate	Uranium Resource estimate (millions of short tons)	Total energy content of resources (quads)	Total energy recoverable (quads)		
			LWRs		LMFBR
			No recycle	Uranium and plutonium	
Low case	1.0	59,000	390	610	41,000
Middle case	1.76	103,000	680	1070	72,000
High case	3.78	222,000	1450	2280	154,000

The uranium resources listed in Table 42 include only those recoverable at prices low enough to justify their use in LWR's. In breeders, however, lower grade (higher cost) resources could be used without raising total generating costs appreciably, since breeder fuel costs are a very small fraction of total generating cost. Thus, resources potentially available for use in breeders would be estimated as considerably larger than those shown in the table. To ensure that nuclear energy continues to play a major role until alternative energy sources become available, it will be necessary to develop and deploy advanced reactor concepts with more efficient fuel cycles. The time when there will be a need for advanced reactors depends on the uranium resources available (and their producibility), the rate of growth in demand for electricity, and the cost of producing energy with such reactors. If the quantity of economically recoverable resources is high--say 4 to 5 million tons of U_3O_8--the required schedule for developing advanced reactors is not critical, meaning that the decision on which concept to adopt can be deferred for up to 10 years. Similarly, if the growth rate of demand for electricity is low--1 or 2 percent per year--then LWR's can meet the need for nuclear power well into the next century. Since it is not possible to be certain about these matters, the panel feels that neither a large uranium resource base nor a low electricity demand growth rate should be adopted as a basis for planning at this time and that development of advanced reactor concepts should be pursued on a priority basis.

The amount of energy that can be extracted from a given resource base varies enormously from one reactor type to the next. Table 42 illustrates the amount of energy that could be extracted by typical LWR's and breeders based on the Uranium Resource Group's low, middle (recommended), and high resource base estimates. The table illustrates the successive gains in energy recovery that can be achieved by recycling uranium and plutonium in LWR's, employing advanced converters, and using fast breeder reactors. Breeder reactors (specifically the liquid metal fast breeder reactor, and the gas-cooled fast reactor), could recover about 70 times as much energy as LWR's from a given amount of natural uranium, and meet United States needs for electricity for centuries. Even the depleted uranium tails stored at the nation's three uranium enrichment plants could provide 12,000 quads of energy if used in breeder reactors. This exceeds the combined total of domestic coal, oil, and natural gas reserves (about 6,500 quads). Thus, even without further uranium mining, enrichment plant tails alone could provide 160 years of domestic energy requirements at current rates of use. <u>It is important to note that these tails are a major fuel source for fast breeders, but not for LWR's or advanced converters</u>, for using them in LWR's or advanced converters would require further enrichment to concentrate the small quantities of uranium-235. Even if this should some day become economically attractive, only a small fraction of the energy content could be recovered.

Improved Light Water Reactor Technologies

In the United States, virtually all nuclear power plants with operating licenses are LWR's, which constitute about 50 gigawatts in generating capacity and provide approximately 12 percent of the nation's electricity. LWR's with a generating capacity of approximately 140 gigawatts are expected to provide about 27 percent of the total electricity supplied in 1987, and this nuclear growth is expected to continue to increase to about 210 gigawatts by the year 2000 (National Electric, 1978). It is evident, therefore, that LWR's will remain the dominant source of nuclear energy in the United States through the year 2010, regardless of current efforts at developing advanced nuclear systems.

As a result of the constraints on fuel recycling imposed by the federal government, LWR's are currently operating on the once-through fuel cycle, which is an inefficient use of this reactor technology. Although these plants are still competitive with fossil-fired plants, their lifetime fuel requirements constitute an excessive consumption of limited uranium resources (Crawford, 1977). Accordingly, there are strong incentives to improve the performance of existing and future LWR plants by increasing the efficiency of fuel use. For example, studies indicate that PWR performance could be improved, mainly through core redesign, to yield a 12-percent decrease in uranium consumption, realized from a 67-percent increase in fuel exposure, as shown in Table 43 (Dietrich, 1978). Recycling of uranium and plutonium alone could reduce current lifetime yellowcake requirements (6,130 short tons U_3O_8 per 1,000-megawatt plant) by 32 percent, to approximately 4,200 short tons.

The 16-percent improvement in PWR uranium use that can be achieved if the simple thorium fuel cycle is substituted for the uranium-plutonium fuel cycle is rather modest if the extensive development effort to establish a commercial thorium fuel cycle industry is taken into consideration; an even smaller benefit in the long run would be realized with the denatured thorium fuel cycle. In fact, any of these strategies would require substantial time and effort to implement and even more time and effort to yield improvements. It is clear that these improvements would not be great enough to resolve current uncertainties about future nuclear fuel supplies satisfactorily.

Advanced Converters

Advanced converters offer the potential for significantly extending the life of our domestic uranium resources, and permit the use of thorium as a fertile material as well. Interest in these reactors has heightened in the past few years as the limitations on uranium availability have become more apparent. The question is whether advanced converters should be developed as an interim step, before introduction of breeders.

There is little doubt that a significant reduction in a reactor's lifetime uranium commitments can be achieved with any of several advanced converter reactors. The lifetime commitments of uranium for these reactors typically will fall in the range of 1,500 to 3,000 short tons of U_3O_8, as compared to roughly 6,000 for LWR's (Figure 21).

An often unrecognized characteristic of advanced converters is that as the conversion ratio is increased, a point is reached at which the reactor requires significantly more uranium in its initial fuel loading than does an LWR. Only after 5 to 10 years of operation (depending on the conversion ratio) will such a reactor produce a net reduction in cumulative uranium requirements. Highly efficient advanced converters therefore will extend the resource base only if they can be deployed soon enough to achieve major savings in uranium before low-cost uranium resources are exhausted. Calculations performed originally for this panel and reported in a recent paper (Perry, 1977), projected uranium requirements for four hypothetical nuclear power growth rates and for four different combinations of LWR's and advanced converters. The assumed capacities were 200, 300, 500, and 800 gigawatts by the year 2000, increasing thereafter by 5, 10, 30, and 50 gigawatts per year, respectively. It was assumed that these power plants would operate at 70-percent capacity factors with enrichment plant tails' concentrations held constant at 0.2 percent uranium-235. The four combinations of reactors and their fuel cycles were:

- LWR's alone, without uranium and plutonium recycling

- LWR's alone, with recycling

- LWR's in combination with high-temperature gas-cooled reactors (conversion ratio 0.76), introduced in 1990

- LWR's in combination with CANDU reactors running on the thorium-uranium fuel cycle (conversion ratio 0.96), introduced in 1990. (Note that this example would require very little uranium after the initial loading.)

Table 43 Pressurized water reactor fuel cycle improvements

Fuel cycle	Design features	Lifetime U_3O_8 Requirements (short tons per gigawatt)[a]
Once-through, uranium	Current pressurized water reactor design; annual refueling; 30.4 megawatt-days per burnup.	6130
Once-through, uranium	Annual refueling; 50.7 megawatt-days per burnup.	5390
Uranium and plutonium recycle	--	4190
Thorium (simple)	U^{235}/U^{233} recycle	3530
Thorium (denatured)	Pu/U^{233} recycle; safeguarded fuel cycle.	3730

[a] 75-percent capacity factor, 0.2-percent enrichment tails, 30 year plant lifetime

Examining each of these cases in turn, we find that in the 200-gigawatt case, advanced converters proved very effective in reducing cumulative uranium demand, thus permitting nuclear power plants to continue operation until the middle of the next century. For this low-growth case, however, LWR's alone, with plutonium and uranium recycling, could continue operating until 2020 without exceeding the 1.8 million tons of U_3O_8 estimated by the Uranium Resource Group in the $30-per-pound forward cost category, a delay of nearly 20 years from the throwaway cycle.

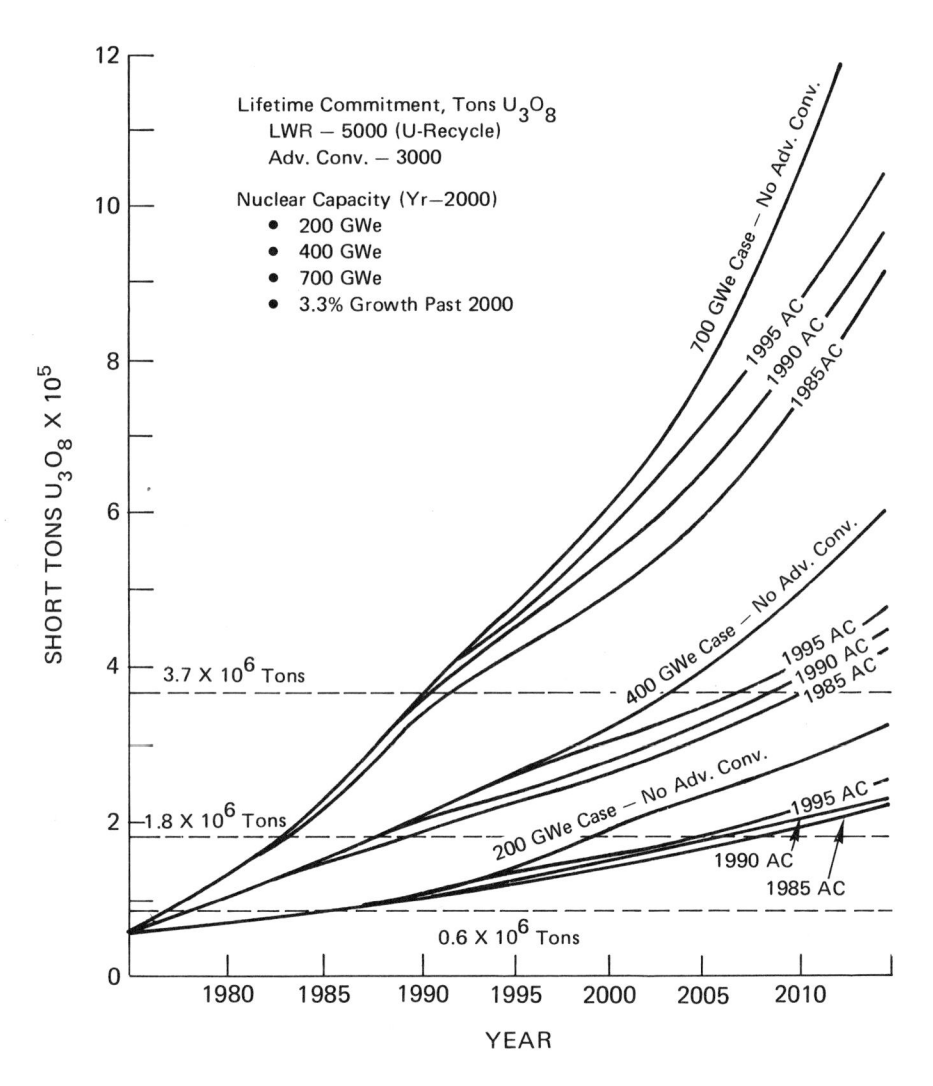

Figure 21 Effect of advanced converter introduction on uranium commitments, in short tons of $U_3O_8 \times 10^5$.

The introduction of advanced converters would quickly dominate the resource requirement picture and could carry the system for a very long time. However, in this case, there would be little market for advanced converters--5 gigawatts of annual capacity additions plus the gradual replacement of retiring LWR's. This market would likely be too small to justify the necessary multibillion-dollar capital investment in the industrial base to manufacture advanced converters, particularly if fuel efficiency is the only motivating force.

In the 300-gigawatt case, with nuclear capacity growing at 10 gigawatts per year, the cumulative uranium consumption values are shown in Table 44, including the lifetime commitment for all reactors ordered through the years 2000 and 2010. The table illustrates that for the most readily available technologies, i.e., fuel recycle and advanced converters with a 0.76 conversion ratio, that uranium consumption on the order of 1.8 million tons is only slightly affected by the introduction of advanced converters. However, if the uranium resource total is as high as 3.8 million tons, advanced converters would offer a significant delay (several decades) before reaching that level. This 300 gigawatt case favors advanced converters only marginally (at a resource of 1.8 million tons of U_3O_8) and it is probably already too late to develop and deploy advanced converters at the rate assumed.

Table 44 Uranium commitments for several reactor combinations in the 300-gigawatt range, in millions of tons

Reactor combination	U_3O_8 2000	Comulative requirement 2010
LWRs with throwaway cycle	2.2	3.2
LWRs with U and Pu recycle	1.4	2.0
LWRs with recycle plus advanced converters (conversion ratio 0.76)	1.2	1.7
LWRs with recycle plus advanced converters (conversion ratio 0.96)	1.1	1.4

In a case intermediate to these, say 400 gigawatts in the year 2000 (see Table 45), again it is shown that advanced converters have only a small effect on the time to consume 1.8 million tons of U_3O_8 but a significant effect on the time for 3.6 million tons.

Table 45 Uranium commitments for several reactor combinations in the 400-gigawatt range, in millions of tons

Reactor combination	U_3O_8 2000	Cumulative requirement 2010
LWRs with throwaway cycle	3.0	4.7
LWRs with U and Pu recycle	1.9	3.0
LWRs with recycle plus advanced converters (conversion ratio 0.76)	1.8	2.4
LWRs with recycle plus advanced converters (conversion ratio 0.96)	1.6	2.2

For installed capacities of 500 gigawatts or more in the year 2000 (see Table 46), advanced converters would have a negligible effect on the time when the cumulative demand for uranium would exceed 1.8 million tons, and would only modestly affect the date when 3.8 million tons would be reached. Thus, even if uranium supplies are found to be double the Uranium Resource Group's prudent planning figure, breeder reactors would have to be introduced before the turn of the century if nuclear generating capacity were growing toward 500 gigawatts or more by then. This is not to say that a particular advanced converter might not have other useful characteristics that would justify its use, only that in regard to their limiting uranium consumption it would be a case of too little too late.

Thus, the aforementioned study indicates that advanced converters would significantly increase the amount of energy generated only under the following conditions: (1) relatively low nuclear capacity growth rates, especially if they show a tendency to level off, (2) introduction of substantial numbers of advanced converters by 1990 (an improbable achievement considering that such reactors would have to be ordered now to be available then), or (3) a uranium resource base significantly larger than currently envisioned. Kasten et al. (1977) analyzed seven advanced converter alternatives to determine their effects on the lifetimes of various sizes of uranium resources; they reached a similar conclusion.

Based on our present knowledge of uranium resources, the probable need for growth of nuclear power, and uncertainties about the success of competing technologies, this panel believes it prudent to proceed without delay to develop and demonstrate fast breeder technology so

Table 46 Uranium commitments for several reactor combinations in the 500-gigawatt range, in millions of tons

Reactor combination	U_3O_8 2000	Cumulative requirement 2010
LWRs with throwaway cycle	3.9	6.1
LWRs with U and Pu recycle	2.3	3.9
LWRs with recycle plus advanced converters (conversion ratio 0.76)	2.3	3.2
LWRs with recycle plus advanced converters (conversion ratio 0.96)	2.2	3.0

that it will be available for wide deployment if needed. If uranium resources, recoverable at costs suitable for LWR's, do prove to be limited to about 1.8 million tons of U_3O_8 and electricity growth rates greater than 4 percent per year are realized, the United States breeder program would appear to be already behind schedule. While a commitment to deployment of breeders at this time is not required, effective national policies must be enacted to accelerate the schedule for breeder technology development.

If the lower growth experience should continue, there might indeed be time, in principle, to put in place a system largely based on advanced converters and to exploit their superior fuel efficiency. Also, while advanced converters do not offer all the potential advantages of breeders and may not delay the date required for breeder introduction, they could be used very effectively in combination with breeders. Once enough breeder capacity is installed to produce a surplus of plutonium, future demand for electricity can be met by installing combinations of breeders and advanced converters, with the breeders fueling the converters. It might prove desirable to use a thorium blanket in such breeders, generating uranium-233 rather than plutonium. Such a combination of breeders and advanced converters would be more economical than the use of breeders alone.

Although there appears to be no need for accelerated development of advanced converters now, continuing to support the orderly evolution of promising concepts on a normal schedule is warranted. The panel recommends that development and demonstration of HTGR technology be continued so that it will be available. Among the potential advanced converter alternatives, HTGR (operating on a gas turbine cycle) appears to be a preferred technology because of its high level of efficiency plus ultimate independence of fresh water cooling.

Thermal Breeder Reactors

Two thermal breeder reactor concepts, the Light Water Breeder Reactor and the Molten Salt Breeder Reactor (MSBR), have received significant attention. These systems both use the thorium fuel cycle and have breeding ratios close to unity (1.0). While this conversion ratio allows some incremental increase over the advance converter concepts the reactors produce no excess fuel to supply other reactors and they do not significantly offer any increase in extending the uranium resource base. However, there is a possibility that the MSBR could be optimized to produce a breeding ratio as high as 1.07, which, because of its high specific power, could give rise to doubling time under 20 years. Because of the close balance in both the neutronic and chemical process accounting, the achievement of meaningful breeding with MSBR has to be regarded as somewhat more speculative than for fast reactors.

Fast Breeder Reactors

Thermal breeders, with their marginal breeding gain, may be barely able to sustain their own fuel needs. A fast breeder reactor, with its high breeding ratio, can produce enough fuel for itself as well as for additional new reactors. Breeder reactor concepts falling under these two basic categories have been under investigation in the United States and other industrialized countries for several decades. Following extensive technical review, a decision was made by the U.S. Atomic Energy Commission in 1967 to assign development priority to the Liquid Metal Fast Breeder Reactor (LMFBR) whose need, proven feasibility, and predicted performance had achieved broad-based support (Ramey, 1968; Shaw, 1966, 1968; U.S. Atomic Energy Commission, 1967).

It should be noted that the performance potential of the gas-cooled fast breeder reactor (GCFR) is virtually the same as that of the LMFBR. In addition, the GCFR has long had the interest of the government, utilities, and the nuclear industry as an alternative to the LMFBR. However, the development effort required for a helium-cooled system was generally judged to be more extensive than for a liquid-metal-cooled system, and as a result, funding for gas-cooled fast reactor studies has been so modest that the development status of this system is many years behind the LMFBR. On a longer time scale the MSBR might be an attractive system complementary to the LMFBR, since it provides a very efficient means for using thorium, and has the lowest fissile inventory per unit power output of any reactor concept. It might be deployed on a time scale comparable to fusion, and is perhaps best considered as an economic competitor to fusion.

The major industrialized countries of the world have all selected the LMFBR to meet their future needs for nuclear energy--which is viewed by the panel as a solid endorsement of the direction taken by the United States in its breeder development program. In fact, Britain, France, and the Soviet Union already have LMFBR's generating electricity on line, while Japan and West Germany have plants in design or under construction. In July 1977, a significant step toward world commercialization of the LMFBR was made when France, West Germany, Italy, Belgium, and the Netherlands signed agreements for future development and marketing of the French Super Phenix design. The first plant of this type is already under construction and scheduled for operation in 1983. It is important to note that these countries have reaffirmed their commitments to developing and deploying the LMFBR despite recent United States policy to the contrary.

LMFBR DEVELOPMENT AND DEPLOYMENT

LMFBR Development Program

Because of their superior performance characteristics, fast breeders have been and continue to be emphasized in United States and foreign breeder development programs. In the United States, the LMFBR program evolved from a number of breeder reactor experiments. The Experimental Breeder Reactor (EBR-1) generated the world's first nuclear electricity in 1951. After this reactor came EBR-2, which is still in operation; the Enrico Fermi 200-megawatt-thermal fast breeder demonstration reactor (constructed in 1968, operated, suffered a partial meltdown, repaired and operated again, and shut down in 1973); and the Southwest Experimental Fast Oxide Reactor (SEFOR) (started up in 1969 and shut down in 1977) with a design to demonstrate the safety of the LMFBR core with regard to power transients. Nearing completion is the Fast Flux Test Facility (FFTF), a 400-megawatt-thermal test reactor to be used in testing fuels, materials, design parameters, and other aspects of LMFBR technology. This facility, authorized in 1967, is expected to achieve full-power operation in 1980.

In the past few years, LMFBR technology has advanced beyond basic feasibility to the extent that prototype pumps, valves, heat exchangers, and other components have been built, tested, and placed in service in large demonstration plants throughout the world. Although a major engineering effort is necessary to demonstrate and deploy any energy concept, no technological breakthroughs are required for LMFBR. Remaining uncertainties have to do mainly with putting the existing technology into commercial practice by designing, constructing, operating, and maintaining commercial-scale units that can compete with other power plant concepts for the 21st century and beyond.

The objective of the national breeder program is to develop a total breeder system to a point that will enable government, utility, and industry leaders to evaluate the potential energy supply role of breeder reactors and to deploy such a system if and when it is needed. As previously noted, progress in foreign and domestic breeder programs continues to support the LMFBR as the most promising breeder concept. Issues related to nuclear proliferation, safety, and economics have been raised against the LMFBR but this panel believes these objections can be satisfactorily resolved without compromising the viability of the LMFBR system.

Development of the LMFBR has now reached the stage at which the various scientific and engineering technologies employed in the LMFBR must be integrated into a total plant system. This can be accomplished only by means of an intermediate size demonstration plant operating on a utility grid that would provide vital and unique information on LMFBR performance and safety. Without such information, the potential for the LMFBR to be licensed and to operate commercially would remain uncertain, and this promising source of abundant nuclear energy would be excluded from the limited inventory of future United States energy supplies. In spite of some opposition to construction and operation of an LMFBR demonstration plant, on the grounds that it would be an irrevocable step toward deployment of commercial plants, it is obvious that this high-technology system would not be deployed if the demonstration plant provided no assurances of commercial practicality.

The Clinch River Breeder Reactor project (380 megawatts-electric) was authorized in 1971 as a major step toward commercialization of the LMFBR in the United States. Plans until very recently called for this reactor to be followed in about 5 years with a full-scale Prototype Large Breeder Reactor, and then by one or more commerical reactors. The Clinch River project has been delayed by the federal government, and the decision whether the Clinch River Breeder Reactor is acceptable as the LMFBR demonstration plant must be made in another forum. However, it should be noted that if the LMFBR demonstration project were allowed to resume at its original pace, operating data would not become available until after the mid-1980's and the earliest startup of the first commercial-size LMFBR could not occur before the mid-1990's.

Numerous studies by the government (Till, Chang, and Rudolph, 1978) and nuclear industry (Westinghouse, 1978) in the United States have established that the fast breeder reactor can operate on the thorium fuel cycle, although less efficiently than on the uranium-plutonium fuel cycle. In particular, it has been determined that potentially acceptable performance can be obtained from a plutonium-fueled LMFBR with a thorium blanket or an LMFBR core with plutonium-thorium fuel and a thorium blanket. Extensive development effort will be required, however, to establish a commercial thorium fuel cycle, even if a thorium-based fast breeder is demonstrated to be technically and economically feasible. In any case, the LMFBR appears to have the flexibility in its fuel cycle to keep it compatible with a modified LWR fuel cycle, such as one based on denatured (uranium-233/uranium-238) fuel. The United States breeder program should continue to investigate the use of the thorium fuel cycle in the LMFBR in support of near-term international efforts at identifying an acceptable proliferation-resistant nuclear fuel cycle.

International Cooperation

The advanced status of LMFBR development in the major industrialized countries of the world has already been described. There are obvious advantages to sharing the results of national LMFBR programs between countries and existing technology agreements should be continued and expanded where practicable. However, complete reliance on foreign breeder technology, which might become a major source of future energy supply, entails risks and uncertainties similar to those associated with the nation's current dependence on foreign oil. It appears that the best policy for the United States with regard to LMFBR development is to continue an independent national effort along with close cooperation and support of foreign programs and needs.

Timing of LMFBR Deployment

The nuclear generating capacity that will be on line in the year 2000 has been projected by the U.S. Department of Energy (1978c) to fall in the range of 250 to 400 gigawatts. Electric utilities, however, estimate that 350 gigawatts will be supplied by nuclear power by that time. The Uranium Resource Group estimates a 33 percent probability that United States uranium reserves and potential resources at a forward cost of $30-per-pound ore will be less than 1.76 million tons. This will probably not be adequate to satisfy the lifetime fuel requirements of LWR's, using the once-through fuel cycle (6,000 tons per plant) that will be on line at the end of the century. The result would be constraints on nuclear generation capacity growth unless more efficient use is made of limited uranium supplies.

Although the LMFBR has reached an advanced stage of development as a result of more than two decades of intensive research and development, considerable effort must be made before this system is available for deployment on a utility grid. Therefore, it appears that the apparently long time that would elapse before utilities could experience nuclear fuel supply problems is actually very short in terms of the time required to achieve the most promising and effective solution—development and deployment of the LMFBR (Figure 22).

LMFBR Economics

In view of the relatively high capital costs of the LMFBR demonstration plants that are in planning, construction, or operating stages at the present time, some concerns have been raised with regard to the economics of the LMFBR. As a matter of fact, the LMFBR entails sophisticated technologies and design features that are expected to result in a higher capital cost than that of a comparably sized LWR. However, the cost of demonstration or prototype plants for any new technology such as the fast breeder reactor cannot be considered as representative of subsequent plants produced by a mature industry. Cost reductions derived from design changes evolving from construction and operating experience are expected eventually to lead to LMFBR capital costs only about 25 percent greater than LWR's. The ability of the LMFBR to be competitive with conventional nuclear power plants and other sources of electric energy will depend on its low fuel cycle costs. Numerous studies have been performed on the economic competitiveness of commercial breeder reactors; the wide range of conclusions is attributable largely to varying assumptions about energy demand, resource availability, and the general economy.

A recent government analysis (Till, Chang, and Rudolph, 1978) of LMFBR economics included an evaluation of the discounted system cost of a mix of LMFBR and LWR power plants as a function of the difference in capital cost between the LMFBR and LWR. As indicated in Figure 23, substantial benefits are realized at the projected 25 percent difference. In fact, the fast breeder appears to be economically attractive at capital costs approaching twice that of LWR's. A major variable affecting the allowable capital cost difference is the price of uranium. Figure 24, from this same analysis, indicates that the capital cost of an LMFBR commercially introduced in the year 2000 can range from 50 to 150 percent greater than that of an LWR, depending on uranium price. The $50-per-pound value at the 5.6 million ton consumption level that defines the low uranium price schedule used in this analysis is closely comparable to the current market price of uranium ($43 per pound) for immediate delivery.

Although other factors will influence the economic competitiveness of the LMFBR, the trend of those parameters affecting its commercial viability appears to justify an aggressive development program to make this advanced nuclear system available if and when it is needed.

LMFBR as Complement to LWR

The basic nuclear characteristics of plutonium clearly establish the uranium-plutonium fuel cycle as the most efficient fuel system for fast breeder reactors. LMFBR designs based on current technology yield a breeding ratio of approximately 1.4 and a compound fuel system doubling time well under 20 years. Development of advanced carbide fuel materials and other reactor improvements could further improve the performance characteristics of plutonium-fueled LMFBR plants.

The uranium-plutonium fuel cycle has been in use in this century and abroad for approximately 30 years, initially for production of nuclear weapons materials by government facilities and subsequently to support civil nuclear power programs. In the United States, LWR's using slightly enriched uranium fuel have been deployed by electric utilities and the nuclear industry on the assumption that the plutonium by-product would be recycled in LWR's and breeder reactors. Aqueous reprocessing of commerical uranium fuel from thermal reactors is already being accomplished in France at Cap La Hague, and reprocessing of research fast reactor fuel was demonstrated in EBR-2 facilities in the United States and at Dounreay in Great Britain. The British facility has recently

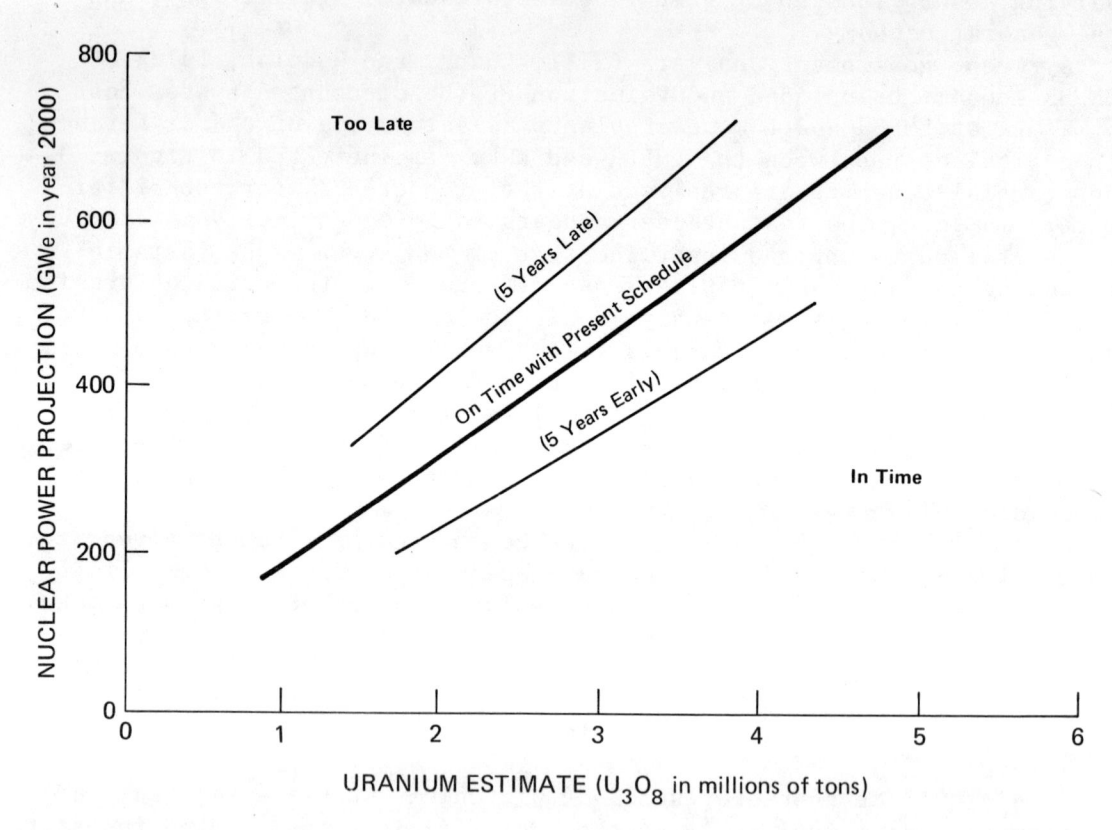

Figure 22 Critical timing for the development of the liquid metal fast breeder reactor. ("Present schedule" implies a 1995 startup date for the first commercial breeder reactor.)

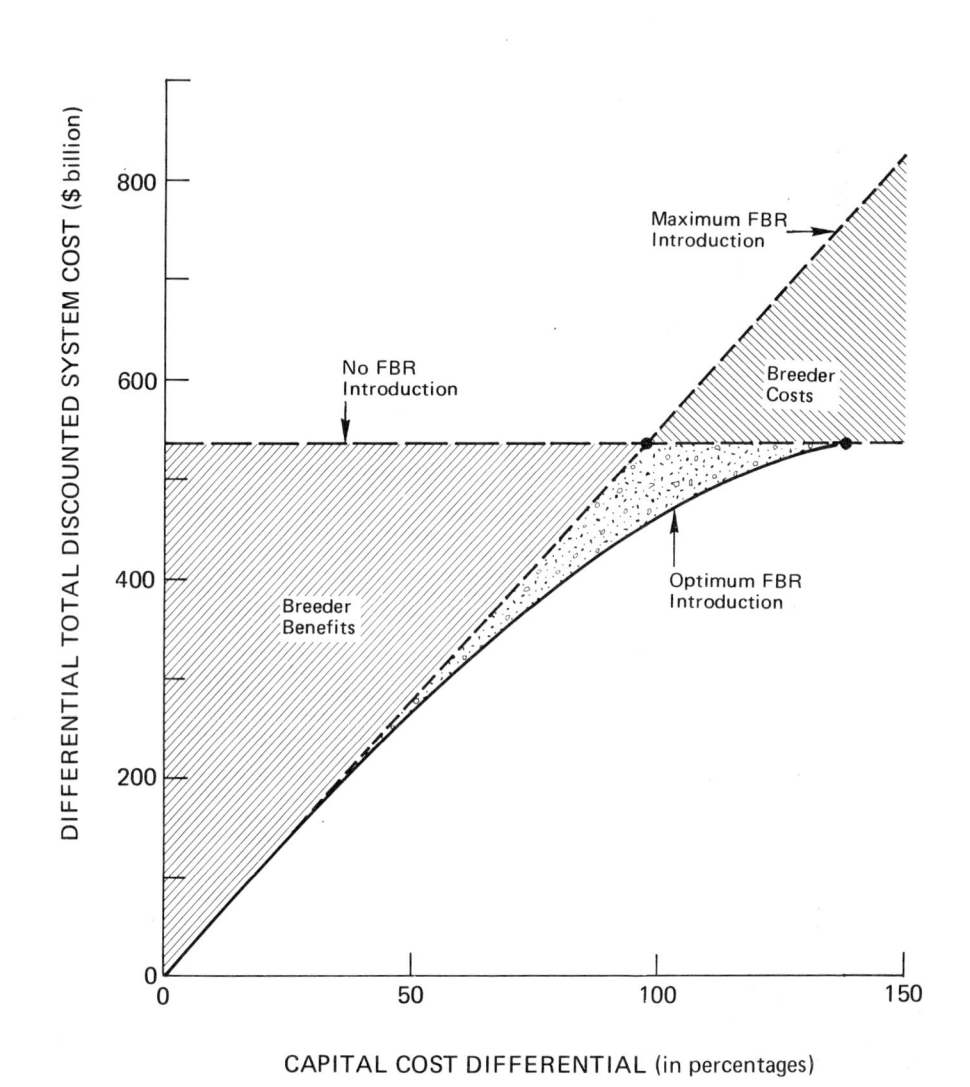

Figure 23 Relationship between discounted cost of nuclear power generation system and the differential costs of breeders and light water reactors, for various breeder deployment scenarios (Till, Chang, and Rudolph, 1978).

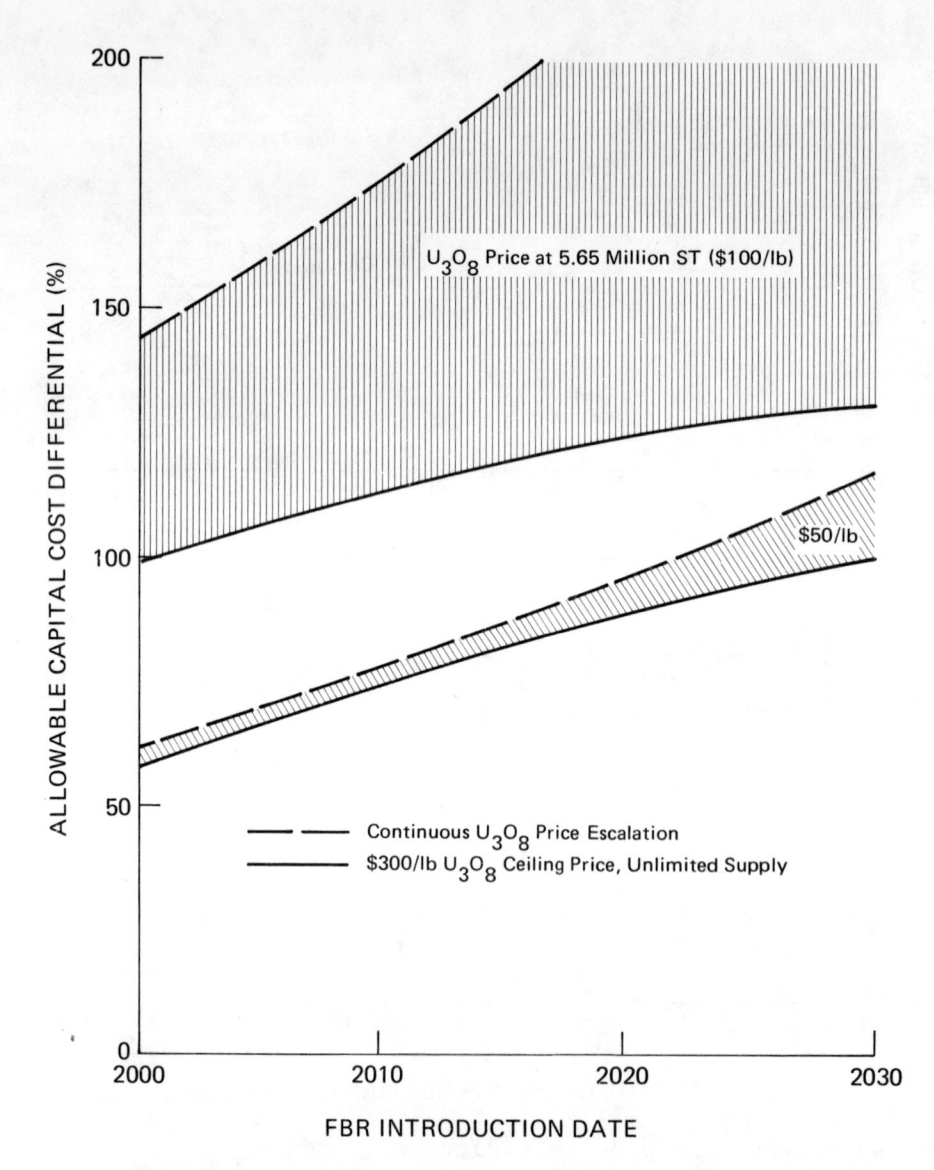

Figure 24 Effect of uranium price on allowable capital cost differential between breeders and light water reactors, at various fast breeder introduction dates (Till, Chang, and Rudolph, 1978).

been modified to handle high-burnup fast reactor fuel and will begin reprocessing uranium-plutonium fuel from the 600 megawatt-thermal Prototype Fast Reactor (PFR) in early 1979. Comparable experience exists in the design and fabrication of uranium-plutonium mixed oxide fuels for fast reactors (Fast Flux Test Facility, Phenix). Therefore, the technology of the uranium-plutonium fuel cycle is well known and requires only modest development efforts to adapt LWR-based processes to high-burnup breeder fuels.

When and if fast breeder reactor plutonium is introduced commercially, fuel will initially be obtained from an external source until the total breeder fuel cycle is able to generate enough plutonium to become self-sustaining. The plutonium produced in LWR's, which will still be the dominant source of nuclear energy in the United States at the turn of the century, will be a practical and economical source of breeder fuel. This arrangement should reduce uncertainties with regard to the long-term availability of uranium as these uncertainties are inhibiting the required increase in nuclear generating capacity. The plutonium would assure electric utilities of an adequate supply of nuclear fuel at stable prices, and hence, the plutonium-fueled LMFBR would complement the LWR fuel cycle and permit an orderly transition to the more efficient fast breeder reactor system of the future. A combination of the technologies would provide utilities with the diversity necessary to minimize the impact of disruptions in fuel supplies, and to control costs through competition.

NUCLEAR POWER GROWTH SCENARIOS

To meet future growth in electricity demand and to displace existing generating capacity currently fueled with oil and natural gas, the Supply and Delivery Panel believes that the United States must rely primarily on nuclear and coal-fired power plants until some time well beyond the turn of the century. A combination of these technologies would be safer, less damaging to the environment, and more secure against interruptions of supply than the use of either source alone. Although the nuclear industry and the utilities should be fully capable of meeting the demand for additional capacity under normal circumstances, they are currently faced with a number of regulatory, institutional, and political problems--national and international--that, unless resolved, could deny the United States the assurance of adequate supplies of electricity. To illustrate how these policies and practices might affect the future growth of nuclear generating capacity, the panel formulated three alternate scenarios reflecting various levels of national commitment to nuclear power.

As estimated by the Uranium Resource Group (National Research Council, 1978) the uranium resource base of 1.8 million tons U_3O_8 is assumed for all scenarios, although a different degree of industry commitment to near-term uranium production is assumed for each scenario. The specific assumptions regarding the actions necessary to achieve these production rates are found in the Uranium Resource Group Report.

For reasons outlined previously, the only advanced reactor type projected to achieve widescale deployment before the year 2010 is the LMFBR.

Business-as-Usual Scenario

This scenario assumes that conditions and practices that now exist for nuclear power will continue indefinitely without appreciable change, namely:

1. No significant improvement will occur in licensing and regulation, with current trends continuing toward increasing complex practices at both the state and federal levels.

2. The host of uncertainties facing utilities in financing, licensing, and constructing nuclear power plants and fuel cycle facilities will continue, severely limiting domestic orders for new nuclear generating capacity.

3. Reprocessing and recycling of unburned uranium and bred plutonium from spent reactor fuels will continue to be prohibited in the United States, limiting the operation of domestic LWR's to a once-through fuel cycle.

4. Utilities will require reasonable assurance of a lifetime supply of uranium before making a commitment to construct new reactors.

5. All domestic nuclear power stations now on order will be completed, but further schedule slippages will occur, thereby reducing the rate of growth of installed nuclear generating capacity.

6. Development work on LMFBR technology will continue, but work on all other advanced reactor concepts will be discontinued owing to a lack of market potential.

7. Introduction of LMFBR's in the United States will be prevented by the failure to design and construct the necessary demonstration reactors and prototype reactors on a timely basis, and by the unavailability of plutonium because of the ban on fuel reprocessing.

8. The number of firms supplying the domestic nuclear industry will drop dramatically.

9. The cost of new nuclear power plants and fuel for existing plants will rise as a result of the limited number of new orders, reduced competition owing to fewer sources of supply,

lack of incentives to develop and implement improvements in technology, and higher uranium prices resulting from failure to reprocess and recycle uranium in spent fuel.

10. The ability of the United States to influence the nuclear practices of other countries will continue to decline, and many nations will move toward achieving total fuel supply independence; nuclear exports of reactors and fuel cycle services will decline as the United States is viewed increasingly as an unreliable supplier.

Given this scenario, installed nuclear power generating capacity (in gigawatts-electric) is estimated as follows:

	1990	2000	2010
LWR	165	210	260
LMFBR	0	0	0
Total	165	210	260

Enhanced Supply Scenario

This scenario assumes that the government will adopt policies and practices to reduce institutional and political uncertainties in nuclear power to the degree that the utilities again view nuclear power as a viable alternative for additional new generative capacity. Specific assumptions include:

1. Positive steps will be taken to streamline the regulatory process, resulting in a substantial reduction in reactor licensing times.

2. A practical solution to safeguarding nuclear materials against proliferation will be developed, and the government will make a commitment to the licensing and operation of facilities for reprocessing spent LWR and LMFBR fuels.

3. All uranium and most of the plutonium recovered from spent LWR fuels will be recycled in LWR's except for that plutonium needed to fuel breeder reactors.

4. The government will proceed promptly with the construction and operation of improved methods and facilities for radioactive waste disposal.

5. The financial community will consider investment in nuclear power plants and nuclear fuel cycle facilities acceptable

risks, and will provide the financing necessary to assure future growth.

6. Nuclear power plants now on order will be completed on schedule; utilities will place some orders for LWR's without an assured reactor-lifetime (30-year) supply of uranium.

7. The breeder development and deployment program will be accelerated to recover partly from the current slowdown.

8. The United States will be considered a reliable source of both nuclear power plants and nuclear fuel cycle services, with the nuclear industry again becoming a strong competitor in the export market.

9. Development and demonstration of attractive advanced reactor concepts other than the LMFBR will continue for possible future deployment to use the excess fissile isotope production from LMFBR's.

For these assumptions, the following power levels (in gigawatts-electric) are estimated:

	1990	2000	2010
LWR	220	500	700
LMFBR	0	2	10
Total	220	502	710

National Commitment Scenario

This scenario assumes a strong government commitment to the full scope of the nuclear power program, particularly with respect to rapid deployment of the LMFBR. Its principal provisions would include:

1. A major streamlining of the regulatory procedures, reducing licensing lead times to an absolute minimum for both nuclear power plants and nuclear fuel cycle facilities.

2. LWR's (and later LMFBR's) installed at the capacity of the nuclear industry to supply reactors and fuel.

3. A firm commitment to the demonstration of LMFBR technology by the mid-1980's and the rapid deployment of commercial LMFBR capacity if and when needed.

4. A national commitment to the reprocessing of spent fuels from both LWR's and LMFBR's.

5. All uranium recovered from spent fuels recycled in LWR's, with recovered plutonium reserved exclusively for fueling LMFBR's until the LMFBR becomes self sustaining or until plutonium is shown not to be needed because of larger-than-projected uranium reserves.

6. The uranium resource availability limitation on installed LWR capacity will be partly relieved through improved fuel efficiency, reducing enrichment plant-tails concentration to 0.1 percent uranium-235 and using somewhat higher forward cost uranium; utilities will place some orders for LWR's without an assured reactor-lifetime (30-year) supply of uranium.

7. If necessary, there will be a nationwide training program to ensure an adequate supply of engineers and construction workers to design and build new generating capacity on schedule.

Under these assumptions, the achievable power levels (gigawatts-electric) are estimated as follows:

	1990	2000	2010
LWR	240	540	750
LMFBR	0	10	100
Total	240	550	850

Rather than being precise predictions of future nuclear power growth rates, the above values should be considered indicators of what realistically could be achieved with a full commitment to nuclear power. Of the three scenarios developed, the assumptions of enhanced supply would appear to be an appropriate goal if the United States is to meet its likely future demand for electric power. However, achieving these generating capacities would require a radical change in present government policies on nuclear energy.

Inasmuch as the uncertainties associated with the nuclear industry as a whole have yet to be resolved through consistent government policy, the panel recommends that the national commitment scenario be viewed as a guide to the potential of nuclear power in the future supply and delivery of energy in the United States. Realistically, however, the goals of the enhanced supply scenario would seem to be the most attainable, considering that the policies of the United States government still must be dramatically readjusted if the power levels it projects are to be achieved.

REFERENCES

Atomic Industrial Forum. 1978. Cost Impacts Related to Nuclear Power Plant Project Durations. Prepared by the Subcommittee on Financing the Nuclear Fuel Cycle. Washington, D.C.: Atomic Industrial Forum.

Brandfon, William W. 1978. Comparative Costs for Central Station Electricity Generation. Paper presented at the Atomic Industrial Forum Conference on Energy for Central Station Electricity Generation, Atlanta, April 18. Available from William W. Brandfon, Sargent & Lundy, 55 East Monroe St., Chicago, Ill. 60603.

Crawford, I.W. Donham. 1977. PWR and BWR Light Water Reactor Systems in the U.S.A. and Their Fuel Cycles. Paper presented at International Conference on Nuclear Power and Its Fuel Cycle, Salzburg, Austria, May. New York: International Atomic Energy Agency (IAEA-CN-36/566).

Dietrich, Joseph R. 1978. The Realities and Illusions of Alternate Fuel Cycles. Paper presented at the International Conference on Nuclear Non-Proliferation and Safeguards, Atomic Industrial Forum, New York, October 22-25. Washington, D.C.: Atomic Industrial Forum.

Energy Daily. 1978. Eklund Warns of Rebound from Nonproliferation Drive. 6(201):4.

Finger, H.B., J.A. Lieberman, and M. McNelly. 1976. Assessment of Energy Centers Versus Dispersed Electric Power Generation Facilities. Washington, D.C.: General Electric Co.

Kasten, P.R., et al. 1977. Assessment of the Thorium Cycle in Power Reactors. Oak Ridge, Tenn.: Oak Ridge National Laboratory (ORNL/TM-5565).

National Electric Reliability Council. 1978. Eighth Annual Review--Overall Reliability and Adequacy of the North American Bulk Power Systems. Princeton, N.J.: National Electric Reliability Council.

National Research Council. 1978. Problems of U.S. Uranium Resources and Supply to the Year 2010. Uranium Resource Group, Supply and Delivery Panel, Committee on Nuclear and Alternative Energy Systems. Supporting Paper 1. Washington, D.C.: National Academy of Sciences.

Nuclear Energy Policy Study Group. 1977. Nuclear Power. Issues and Choices. Cambridge, Mass.: Ballinger.

Perry, Alfred M. 1977. Thermal Breeders in Today's Context. Paper presented at the International Scientific Forum on an Acceptable Nuclear Energy Future of the World, Ft. Lauderdale, Fla., November 7-11.

Ramey, J.T. 1968. Utility Participation in the U.S. Atomic Energy Commission Fast Breeder Reactor Program. Paper presented at the American Power Conference, Chicago, April 23.

Shaw, Milton. 1966. Fast Breeder Reactor Program in the United States. Paper presented at the British Nuclear Energy Society Conference on Fast Breeder Reactors, London, May 17-19.

Shaw, Milton. 1968. U.S. Fast Breeder Reactor Program--The Need For and the Status Of. Paper presented to the World Power Conference, Moscow.

Till, C., Y. Chang, and R. Rudolph. 1978. Alternative Fuel Cycle and Deployment Strategies: Their Influence on Long-Term Energy Supply and Resource Usage. Technical source material prepared at Argonne National Laboratory for the U.S. contribution to the International Nuclear Fuel Cycle Evaluation, Working Group 5. Germantown, Md.: U.S. Department of Energy (INFCE/5-TM-2).

U.S. Atomic Energy Commission. 1967. Civilian Nuclear Power: The 1967 Supplement to the 1962 Report to the President. Oak Ridge, Tenn.: U.S. Department of Energy Technical Information Center (TIC-23729).

U.S. Atomic Energy Commission. 1974. Nuclear Power, 1974-2000. Washington, D.C.: U.S. Government Printing Office (WASH-1139 [72]).

U.S. Department of Energy. 1978a. Fission Energy Program of the U.S. Department of Energy, FY 1979. Washington, D.C.: U.S. Department of Energy (DOE/ET-0048 [78]).

U.S. Department of Energy. 1978b. Statistical Data of the Uranium Industry. Springfield, Va.: National Technical Information Service (GJO-100 [78]).

U.S. Department of Energy. 1978c. U.S. Central Station Nuclear Electric Generating Units: Significant Milestones (Status as of April 1, 1978). Springfield, Va.: National Technical Information Service (DOE/ET-0030/2 [78]).

U.S. Energy Research and Development Administration. 1976. Uranium Industry Seminar. Washington, D.C.: U.S. Energy Research and Development Administration (ERDA GJO-108 [76]).

Waste Solidification Panel. 1978. Solidification of High-Level Radioactive Wastes. Report to the Committee on Radioactive Waste Management, Commission on Natural Resources, National Research Council, Washington, D.C.

Westinghouse Electric. 1978. Preconceptual Study of Proliferation Resistant Heterogeneous Oxide Fueled LMFBR Core, Final Report 1978. Advanced Reactor Division. Madison, Pa.: Westinghouse Electric.

Wivstad, Ingvar. 1978. Swedish Utilities Design "Completely Safe" Disposal Route. Nuclear Engineering International 23 (266):45-48.

6 ADVANCED ENERGY SOURCES

Advanced energy sources, for the purposes of this report, are those that can be considered virtually inexhaustible in that they can provide very large amounts of energy for very long periods of time. These energy sources--thermonuclear fusion, solar energy, and geothermal energy--along with nuclear fission using breeders, constitute all the energy supplies now known that can contribute substantially to meeting energy needs for the indefinite future. As the title suggests, these energy sources are not now widely used.

Appraising the potential contributions of advanced sources to the future supply of energy is complicated. For sources accessible by well-developed technologies--such as geothermal dry steam for electricity production or solar energy for water heating--economic competitiveness and the rate of market penetration can probably be predicted now. For systems and applications still requiring technical development, the probability of technical and economic success and the future rate of market penetration can only be estimated. To estimate future potentials, the panel has used a range of assumptions about the probabilities of economic and technologic progress and about possible incentives to assist market penetration. There is likely to be competition among these emerging technologies, and it is therefore incorrect to assume that their potential maximum contributions are additive. Also, the efficiency and cost of most of the solar and geothermal technologies depend on geography and other variables of nature, so that the timing and extent of their commercial use will vary from region to region.

The panel agrees that the United States must shift first from its reliance on gas and oil first to transitional energy sources (mainly coal and nuclear converters) and then ultimately to advanced sources. These changes will take place slowly because they require altered investment patterns for suppliers and users, resolution of technical problems, and turnover in existing energy-using equipment. For these reasons, it is not prudent to depend on the less well quantified promises of the advanced technologies and fail adequately to support the transition sources, with their relatively well-known potentials and problems. To assure adequate energy supply, both types of technologies should be supported.

SOLAR ENERGY

The technologies discussed here cover virtually all the techniques for using solar radiation--whether collected initially in a manufactured collector or by natural processes. For the purpose of this study the technologies have been placed in six categories: heating and cooling for buildings, process heat, bioconversion, solar thermal electricity generation, photovoltaic conversion, and wind energy conversion.

The Solar Resource Group (National Research Council, in press) has considered briefly the potential energy contributions before 2010 from other renewable sources or from other technologies that use solar energy, including ocean waves, tidal power, and space satellites for conversion of solar energy that is then transmitted to the earth. The panel concluded that the probable contributions from these sources do not warrant their inclusion in a study on national energy policy and so they are not among the economic models presented here.

Many studies of the potential of solar technologies have been conducted, and results have varied greatly. Part of the task of the Solar Resource Group has been to understand the reasons for these differences. First, there are uncertainties about the solar technologies themselves; only a few of these studies have had enough available experience to provide a solid basis for cost estimates. Second, there are uncertainties about the costs and availability of competing energy sources. Third, without a national energy policy, there are uncertainties about the financial and institutional environment in which solar energy will be produced and used.

Solar energy has vital characteristics that distinguish it from virtually all the more conventional energy forms. These qualities may affect the manner in which it is evaluated.

- Solar energy is well distributed geographically, but in low concentrations.

- Although solar energy conversion is resource-intensive resulting from the large areas of collection surface required, most of the material resources needed are abundant.

- Although no technology is without some effect on the environment and public health, many solar energy technologies appear more benign and, therefore, less controversial than many conventional ones involved in the public energy debate today.

- In most cases solar energy systems attain optimum performance in modules that are smaller in scale than those common in other energy technologies.

- Methods for using solar energy to heat space and water are easy to understand and use.

- Production of solar energy is intermittent and subject to interruption by the daily solar cycle and inclement weather. Therefore, direct solar systems must have storage capability or a backup system.

- Distributed solar energy systems that use utility-generated electricity for back-up, if not designed to include adequate energy storage, can adversely affect the operations of electric utilities.

The Solar Resource Group evaluated the possible contributions and probable costs of energy produced by the major solar technologies. These estimates cannot be understood without knowledge of the procedures that were used in their development. Since the technologies are quite different and the assumptions about cost vary greatly from one to the next, they are summarized here.

Direct Use of Solar Heat

This category of solar technologies includes all those in which solar heat is collected and used without being converted into a secondary form such as electricity or chemical fuel. These technologies are generally suitable for space heating and cooling, for water heating, and for low-temperature industrial process heat.

Water Heating

Of all solar applications, solar water heating is most likely to achieve early, widespread use. The technology is well developed, and solar water heaters are fairly easy to install and can be used all year long, unlike solar space heating systems. Solar water heating appears economically competitive in some parts of the nation now.

Space Heating and Cooling

The potential for solar space heating cannot be understood without considering the important role of energy conservation in buildings. Energy conservation practices such as insulation, for example, could reduce the need for energy use in buildings by more than half. This could lower the potential energy contribution from a solar heating system, thus raising the cost per Btu of useful solar heat.

Solar heating and cooling methods can be classed as "active" and "passive." Active systems use special collectors, with pumps or fans to circulate air or water to deliver heat where it is needed. Passive systems collect and transfer heat without such mechanical devices, using instead design features of the building, such as heavy insulation, large south-facing windows, and natural ventilation. It is generally agreed that passive methods are simpler and less costly than active ones. Therefore, the thermal design of buildings should be such that conservation measures are implemented, then passive uses of solar energy are incorporated, and finally active solar systems are applied. It was assumed, in the estimates of energy contributions and costs, that these measures gain a market in this order as the costs of fuel and electricity increase. The comparative costs of solar heating and cooling versus fossil fuels or electricity is strongly climate-dependent.

Although our analysis indicates that passive systems offer the least costly approach to space heating, government-sponsored research programs have emphasized active systems. Solar demonstration homes tend to be standard tract homes with high-technology solar heating systems tacked on. The panel endorses the more appropriate approach of designing, constructing, and maintaining buildings that use the natural features of the external environment, especially sunlight, to best advantage. Such a program would stress the design of passive solar systems with heavy emphasis on architectural design. Development of passive systems should be assigned a higher priority in a federal energy policy. Unfortunately, proposed tax credits apply only to active systems, and therefore may have the effect of discouraging more cost-effective measures.

Active solar cooling (air conditioning) for buildings will be among the slowest of solar applications to see widespread use. The technology is poorly developed and expensive, and no significant technical advances appear likely to change the cost picture in the near future.

The major factors that will influence the economics of solar heating and cooling are the interest rate and taxes charged the owner of the system. A tax credit for homeowners who install solar equipment may be justified as a conservation measure in the national interest. Some consumers will probably choose to install a solar energy system for a new building as a form of insurance against rapidly rising fuel costs. The substantial first costs of active systems, however, may deter many, even if the systems are competitive on a Btu-for-Btu, life-cycle-cost basis with conventional systems.

Industrial Process Heat

A substantial opportunity for utilizing solar energy exists in process heat applications. At present, natural gas supplies about half the energy needed for industrial and agricultural processes. It is likely that larger industrial establishments will ultimately shift to coal, but because of handling and storage problems there may be incentives for some industries to convert to solar energy. First, however, the technical feasibility and commercial potential of solar collectors and storage systems must be demonstrated, and their economic competitiveness established (particularly reliability over a reasonable period of time). Solar process heat would be most important in the generation of steam and hot water in industry and of hot air for drying applications in agriculture and food processing. The Solar Resource Group assumed that during the period with which this study concerns itself such systems would be adopted only in sunnier regions.

The proportion of federal research and development funding for solar energy process heat does not reflect its potential for industrial applications. The fiscal year 1978 Department of Energy budget provided less than one-third as much funding for process heat as for residential solar heating and cooling. Industry seems a prime candidate for direct use of solar energy, because life-cycle costing is common for industrial comparison of energy alternatives, the demand is not seasonal, and the need to shift from gas to a more dependable energy source will be an incentive.

There is a strong implication in government planning that technical progress in solar process heat should depend on the progress in heating and cooling technology for residential buildings. This is incorrect, since process heat applications depend on low-cost collectors providing low-pressure steam at temperatures higher than those for space heating and cooling. The Department of Energy should accelerate its research and development on solar process heat and give high priority to the development of collectors and storage systems that are compatible with standard industrial steam systems.

Biomass Conversion

Wood (solar energy captured by vegetation or "biomass") was the major source of energy for human beings until a century ago. In 1875 it provided two-thirds of the energy consumed in the United States. Bioconversion offers many advantages over other methods that employ solar energy entrapment. Convenient storage is especially important. The energy stored in plant tissue and other biomass forms can be used directly by burning, or indirectly after conversion to liquid or gaseous fuels. Of course, indirect conversion is less efficient, but it does offer fuels in forms that are readily usable.

The four major sources of biomass considered here are municipal solid wastes, agricultural residues, plants grown especially on terrestrial energy farms, and plants grown on marine energy farms. Cost estimates were obtained after first considering the total energy content available in each source and then applying a recovery factor.

Estimates for recovery of energy from municipal wastes--based on existing technology--are a reflection of the projection that the volume of such wastes will be reduced by the turn of the century because of political and environmental pressure to recycle all manner of waste. Estimates for agriculture residues are less reliable than those for municipal wastes because of poor data on the availability of residues and the unpredictability of future farming practices. There is evidence that the availability of crop residues has been greatly exaggerated in some earlier studies.

The required technologies exist in agriculture and forestry to produce energy for direct combustion or conversion to other fuel forms by anaerobic digestion. The energy sources are primarily cellulose; plant strains that would offer higher fuel values should be developed, especially those that would yield more lignin-like components, oils, and hydrocarbons. Optimistic energy-farming concepts go back several years, when crop surpluses were common in the United States and there was less competition with food production for land use. The final biomass source considered, marine energy farms, is largely hypothetical and the technologies insufficiently developed so that the estimates given for this source are the least reliable of all.

As noted elsewhere in this report, the most pressing near-term energy problem is to maintain the supply of liquid or gaseous fuels. Accordingly, it appears that a high priority in the solar energy program should be to develop such fuels in forms that can be produced and used efficiently. This, however, has not been the case. The allocation of funds in the Department of Energy's fiscal year 1978 budget showed $17 million for bioconversion, the source of solar-produced fluid fuels, as compared with $223.8 million for solar electric technologies. If solar energy is to contribute to the supply of storable and transportable fuels, the areas already included in the bioconversion program require greater support. Alternative concepts such as hydrogen production, electrolysis of water, and photochemical fuel production should also be supported. Indeed, direct photochemical production of fuels may have the greatest impact of any of these concepts in the long term.

Solar Electric Technologies

Solar technologies for producing electricity are various. They include such relatively well-developed methods as using wind-powered turbines, as well as several concepts that to be useful will require significant technical advances. In the latter category are photovoltaic conversion (so-called solar cells) and--even more speculative--ocean thermal energy conversion, or OTEC, which would exploit the temperature difference between the surface and the deeper waters of tropical oceans to run heat engines and generate electricity. Intermediate in technical complexity, but by no means ready for wide commercial use, are various plans for solar thermal conversion, which involve concentrating solar heat on boilers with arrays of mirrors to generate steam for electricity production.

Solar Thermal Conversion

Solar thermal conversion could produce energy for either large power plants using the central receiver concept, in which solar radiation is focused by large fields of mirrors onto a central boiler, or in smaller, distributed facilities (perhaps in so-called solar total energy systems, in which waste heat from electricity generation would be used for low-temperature applications such as space heating). Major programs sponsored by the Department of Energy and the Electric Power Research Institute are under way to develop the central receiver concept. One area where central receivers could be used is in the Southwest--where numerous 100 megawatt units could provide the Southwest with intermediate load inputs to local utilities. As solar input increases beyond the requirements for intermittent inputs, increased storage capacity will become necessary.

The economics and institutional aspects are more complex, and the type of conversion technology more uncertain, for solar total energy systems. The primary advantage of this method is that reject heat is used for other purposes. The estimates for central station and solar total energy assume improvements in technology, particularly for energy storage, with solar collectors being the main expense for both systems.

One important question in determining the division of support between central station and distributed generation relates to the relative effectiveness of the two concepts. Until recently, the distributed concepts had not been well defined and, therefore, received little support and the central receiver concept has continued to be emphasized. A high priority for government programs should be to define relative benefits and disadvantages of distributed and centralized solar electric systems so that available funding is apportioned wisely.

Photovoltaic Conversion

Although the technical feasibility of photovoltaic solar energy conversion has been demonstrated, high capital costs discourage its use. Thus, reducing costs is a prime goal of current research and development. As As with most other solar technologies, energy storage or a source of auxiliary energy is generally needed for local systems.

It is expected that low-cost photovoltaic systems will be developed in three phases. The first phase--until 1985--will emphasize early system demonstration, with government purchases subsidizing the system until industry capacity is established. In the second phase--to 1995--there should be a shift from current technology to low-cost mass production technology now being developed. The third phase should see a growing industry with slowly increasing production rates and refinements in methods and processes.

The existing Department of Energy program overemphasizes first-phase demonstration projects. The danger here is that production facilities could soon become obsolete as the new technologies needed to reduce photovoltaic costs are developed. Needed, then, is an expanded research program to develop lower-cost technologies. The federal government has not established a broadly based research program on photovoltaics. University groups are so minimally engaged in the program that it does not compare to advanced work being done in this field. The panel recommends that the Department of Energy revise its program to include more research, including increased university participation, and fewer demonstrations of expensive, current technology.

Wind Energy Conversion

Estimated potential and costs of energy from wind depend more on the resource's geographical distribution than those of any other solar technology. Local wind energy resources vary by factors of 10 or more, even between nearby locations, and the best wind sites are often far from points of energy demand. The future cost of wind-driven electricity generation is unknown and depends on such factors as the variability of windspeed, scale of units, and local electricity demand.

Unfortunately, because the wind cannot be turned on at will to meet peaks in demand, a wind generator cannot obviate the need for much generating capacity. Thus, wind-generated electricity can be said to be worth only about 30 mills per kilowatt-hour--(National Research Council, 1978a), the fuel cost today for the gas turbines and diesel engines used in meeting peak loads. The fuel cost for electricity produced by large coal and nuclear base-load plants is, of course, much less.

Considering the many unknown and unpredictable factors, almost all of them nontechnical, estimates of the energy that is likely to be produced from wind amount to little more than educated guesses.

The existing federal program is a rational one that makes good use of available funding, but wind-driven energy will not be ready for commercial use for at least several years. Coexisting with utility systems remains a problem, and no substantial production capacity exists. The panel recommends that the Department of Energy sponsor the development of a dozen or more one-megawatt units so that the design effort can be given to industries experienced in cost-effective design and production. The design program should establish clusters of anemometers to determine the relationship of the resource to utility loads, and should include ways to subsidize the initial development of a wind energy industry.

Ocean Thermal Energy

Ocean thermal energy conversion (OTEC) employs solar energy indirectly to produce large amounts of power from the thermal energy in tropical oceans. No energy need be stored, and the system can operate day and night and throughout the year. However, electricity must then be delivered to shore. One way is by cables installed along the sea floor; another would require that hydrogen be produced at the OTEC station by water electrolysis and then delivered to shore by pipelines.

Cost estimates for ocean thermal plants have been the subject of controversy. The cost of heat exchangers deserves special discussion. Thermal efficiency of an OTEC plant is low because of the limited range of working temperatures. Therefore, the amount of heat to be exchanged is great and the exchangers themselves large. If pessimistic assumptions about each of three design factors--the heat transfer coefficient (including fouling factors) of the heat exchangers, the allowable temperature drop across the heat exchangers, and the cost of the heat exchangers per square foot--prove realistic, the cost of OTEC will be prohibitively high. On the other hand, if more optimistic estimates about these three factors are correct, OTEC may be competitive even at present costs of electricity. Estimated lifetime costs of the heat exchangers must, however, include realistic maintenance and replacement figures.

It is unfortunate that several years were spent on design studies for OTEC plants before the important uncertainties about heat exchangers and other major components were addressed. Fortunately, the budget for the OTEC research and development program has been increased to support laboratory and field experiments. These experiments and subsequent demonstrations should provide data needed to resolve questions about the efficiency of ocean thermal energy.

Solar Energy Scenarios

National Commitment

This scenario is based on the assumption that by 1985 there will be a national policy to foster the use of solar energy, with impetus coming not from the usual economic forces but rather from federal intervention in the market. It is also assumed that after 1990 this policy would require use of solar energy for all new buildings and for all technically practical industrial process heat applications. It would also require reclamation of the energy in municipal and agricultural wastes and schedule the deployment of several solar electric technologies.

Direct use of solar heat. The panel's projections for direct use of solar energy are found in Table 47.

Table 47 Projections for direct use of solar energy--national commitment scenario, in quads per year

Application	1975	1985	1990	2000	2010
Domestic water heating	--	0.2	0.4	1.2	1.3
Passive space heating	--	0.1	0.2	0.3	0.4
Active space heating	--	0.1	0.1	0.6	1.2
Nonresidential air conditioning	--	0	0.1	0.4	1.5
Industrial process heat	--	0.2	0.4	1.6	6.6
Total	--	0.5	1.1	4.1	11.0

Biomass conversion. The panel assumed for the purposes of this scenario that by the year 2000, 95 percent of the nation's municipal wastes will be processed for their energy content, and that 35 percent of the energy content of agricultural residues will be recovered as methane. The results are shown in Table 48. (The possible contributions of energy farms are not included here. The Solar Resource Group estimates that such farms could provide another 3.4 quads by the year 2010.)

Table 48 Projections for energy from biomass conversion--national commitment scenario, in quads per year

Source	1975	1985	1990	2000	2010
Municipal wastes	--	0.5	0.8	1.9	1.9
Agricultural residues	--	0.5	0.9	3.5	3.5
Total	--	1.0	1.7	5.4	5.4

Table 49 Projections for energy from solar electric conversion-- national commitment scenario, in quads per year

Technology	1975	1985	1990	2000	2010
Central station	--	0	0	1.7	8.7
Total energy	--	0	0.1	0.5	1.9
Wind	--	0.1	0.5	1.4	1.8
Total	--	0.1	0.6	3.6	12.4

Solar electric conversion. Several technologies have been developed for converting solar energy to electricity, so it is not possible to predict which ones would be selected for use in a solar-intensive scenario. Therefore, the projections in Table 49 are based on central station solar thermal conversion, solar thermal total energy sytems, and dispersed wind energy systems. (It should be noted that photovoltaic or ocean thermal conversion may one day serve as a conversion technology instead.)

This corresponds to an installed capacity in the year 2010 of 250 gigawatts of central station solar thermal plants (load factor of 0.4), 74 gigawatts of total energy generation (load factor of 0.3), and 50 gigawatts of wind turbines (load factor of 0.4).

Solar energy totals. Combining the aforementioned estimates for total solar energy, we have the results shown in Table 50.

Table 50 Projections for total solar energy supply--national commitment scenario, in quads per year

Application	1975	1985	1990	2000	2010
Direct use	--	0.5	1.1	4.1	11.0
Solar electricity	--	0.1	0.6	3.6	12.4
Biomass	--	1.0	1.7	5.4	5.4
Total	--	1.6	3.4	13.1	28.8

Table 51 Projections for total solar energy supply--enhanced supply scenario, in quads per year

Application	1975	1985	1990	2000	2010
Direct use	--	0.4	0.8	2.2	3.3
Solar electricity	--	0	0.1	1.8	5.5
Biomass	--	0.5	0.8	1.9	1.9
Total	--	0.9	1.7	5.9	10.7

Enhanced Supply

This scenario (Table 51) implies a decision to develop solar technologies in many but not all possible forms. As in the national commitment scenario, market intervention by the federal government is assumed. For heating and cooling technologies, it is assumed that water heating and passive space heating are to be encouraged, and that active space heating and nonresidential air conditioning are not. The figures for industrial process heat are based on the Solar Resource Group estimate for their baseline scenario--a 10 percent growth rate. For bioconversion estimates, it is assumed that only municipal wastes contribute, whereas for solar electricity, the contributions are derived from the Solar Resource Group report baseline for central station and total energy systems and from their baseline with a 10 year delay for wind.

Business as Usual

This scenario depends on the assumption that the costs of other energy sources follow the baseline schedule of the Synthesis Panel's Modeling Resource Group report (National Research Council, 1978a), and that the costs of solar energy technologies follow the estimates of the Solar Resource Group (National Research Council, in press). Under these assumptions, estimates for direct use of solar energy and for solar electricity can be drawn from other reports of the CONAES study. The Demand and Conservation Panel report (National Research Council, 1979) examined the case in which energy costs remain constant in 1975 dollars. In this case, direct use of solar energy remains very slight, reaching only 0.3 quad per year in the year 2010. In scenarios with similar assumptions about the future costs of coal and nuclear electricity, the Modeling Resource Group (National Research Council, 1978a) found that there would be no market penetration for solar electricity by the year 2010. Combining these estimates results in the low projections of Table 52.

CONTROLLED NUCLEAR FUSION

There are two main variants of nuclear energy that offer long-term energy potential: fission and fusion. The prospects for fission energy are described in the chapter on nuclear energy. Fusion denotes a class of rearrangement reactions involving the nuclei of the lighter elements in the periodic table. In these reactions, two charged nuclei approach one another closely with energy high enough to overcome their mutual electrostatic repulsion and fuse, yielding a heavier nucleus, energy, and in some cases a neutron. The resources to fuel the fusion reaction are so plentiful as to offer almost limitless potential energy supplies.

The only way in which fusion has been harnessed thus far is in the hydrogen bomb. A way to exploit the energy from fusion for civilian applications has yet to be developed. Fusion research programs have been directed toward two ways that seem practical for reactor development: the magnetic confinement approach and the inertial confinement

Table 52 Projections for total solar energy supply--business-as-usual scenario, in quads per year

Application	1975	1985	1990	2000	2010
Direct use	0	0	0	0	0.3
Solar electricity	0	0	0	0	0
Biomass	0	0	0	0.1	0.3
Total	0	0	0	0.1	0.6

approach. The long-term goal is to develop fusion to the point that its technical and economic feasibility can be established and it becomes competitive with other long-term possibilities. The panel's Fusion Assessment Resource Group report (National Research Council, 1978b) discusses fusion research in some detail.

The principal application suggested for fusion has been the generation of electricity. Like fission, fusion is likely to be developed in large packages (1,000 megawatts or greater). Since a direct output of a fusion reactor would be energetic neutrons, progressive applications have been suggested. The proposed applications include production of fissile material for fission converters, radiolysis of water to produce hydrogen or of carbon dioxide to produce carbon monoxide as fuel, and transmutation of actinide waste to shorter-lived nuclei for ease of disposal.

Fusion technology must evolve through three stages: scientific feasibility, engineering feasibility, and commercial feasibility. Scientific feasibility requires attainment of reactor-grade plasmas; scaling laws that are well understood; and energy (breakeven the energy released in the reaction equals the energy invested in the plasma). Engineering feasibility implies a demonstration that a suitably designed power-producing reactor can be constructed and successfully operated, with due regard to safety and environmental impact. Commercial feasibility requires a demonstration that reactors of proper design will have all the features necessary to make them potential economic competitors with other commercial energy sources. Although considerable progress has been made, scientific feasibility has not been demonstrated. It is considered likely that scientific feasibility will be established for magnetic confinement, and perhaps for inertial confinement, within the next five years.

The question that remains, however, is whether any of the approaches that promise scientific feasibility in the near future will be appropriate for practical, commercial fusion reactor technology. The principal concerns here are the capital cost, minimum plant output, plant maintenance and availability, plant complexity, and environmental characteristics. It is imperative that the federal government and the ultimate customers, the utilities, cooperate during development of the technology.

The panel believes that a national program should continue to concentrate on the main approaches now being pursued without commitment to any single concept. At the same time, other physics and engineering options must be welcomed and explored in sufficient depth over the next 5 to 10 years to determine their desirability from the user's perspective. The move to pilot-plant experiments should not be attempted until more is understood about confinement, plasma physics, and materials properties. Uses of fusion energy for other than the generation of electricity should be analyzed.

If the fusion program is continued at a high enough level of funding, the panel believes that by about 1990 it may be possible to judge the prospects for starting a commercial demonstration project. When and if such a project is undertaken, it is likely to take about 20 to 25 years and cost about $15 to $20 billion (1975 dollars) to complete a successful demonstration.

Fusion technology is not developed sufficiently to permit comparison with other long-term energy resources, and its development to such a point will undoubtedly be costly. However, other long-term energy systems are not so trouble-free to let fusion remain unexplored.

The panel concludes that with improved scientific understanding and technological advances, achievable under a well-supported government program, fusion will become more attractive as an ultimate long-range contributor to the world energy supply. It is also important to continue the international cooperation that has been so fruitful in this field.

GEOTHERMAL ENERGY

In principle, geothermal energy refers to all heat contained in about 260 billion cubic miles of rocks and metallic alloys at or near their melting temperatures, constituting the entire volume of the earth except for a relatively thin, cool crust. Actually, the practical potential is but a small fraction of the earth's volume in which crustal rocks, sediments, volcanic deposits, water, and steam and other gases at temperatures high enough to be useful are accessible from the earth's surface and from which it may be possible to extract usable heat economcally. Even this, however, is an enormous reservoir of thermal energy, one that will expand to continually greater depths as the technology for recovering and using it improves and the need for it increases.

So far geothermal energy has been used only when it has already been extracted from hot rock by naturally circulating ground-water which has brought it to, or nearly to, the earth's surface in the form of steam or hot water. Of course it also exists in and is potentially recoverable from the rock itself. Geothermal reservoirs are commonly classified into six types, principally on the basis of differences in the media in which the heat exists, their temperatures, and the methods used to recover the heat. Geothermal heat can, at least conceivably, be recovered from hot water reservoirs, from natural steam reservoirs, from brine in so-called geopressured reservoirs, from dry rock heated by the normal geothermal gradient or by local hot spots (the so-called hot dry rock category), and finally bodies of lava, or molten magma. In general, each of these six types of geothermal reservoirs provides access to heat energy. This heat can have a wide variety of uses for which low-grade heat is sufficient, or, if the temperatures are high enough--above about 180°C--it can be used to generate electricity.

The Geothermal Resource Group (National Research Council, 1978c) estimates potential contributions from each geothermal type on assumptions about the individual technologies, the probable costs associated with each, and the institutional and policy considerations affecting their production.

Hot Water Reservoirs

In these reservoirs, water at temperatures up to 350°C or more is trapped underground in permeable formations from which wells drilled down from the surface can recover the water. Because the solubilities of most minerals increase with temperature, the hotter geothermal waters are generally more highly mineralized and therefore cause corrosion, scaling, and waste disposal problems after the useful heat has been removed. These difficulties and the possibility of subsidence of the land surface from which the water is withdrawn have so far prevented large-scale commercial use of this geothermal resource in the United States. The higher-temperature natural brines are, however, being exploited successfully in several other countries. This success must be demonstrated, however, over expected plant operating lifetimes. Hot water reservoirs can be expected to provide a significant fraction of the total future contribution of geothermal energy once demonstration plant success has been achieved.

Natural Steam Reservoirs

Under unusual geological circumstances, the pressure in a hydrothermal reservoir may be low enough for the water to boil and produce steam spontaneously. Except for a variable content of noncondensable gases such as carbon dioxide and hydrogen sulfide, the steam is pure and can be piped directly from the well to a turbine generator, as it is at The Geysers, in California, the world's largest geothermal power development. The steam field at The Geysers and any others that are discovered can be used to produce electricity and therefore will be developed by industry as rapidly as institutional constraints permit. Since natural steam fields are rare, the only effect they will have will be of a local nature and not on the nation's total energy supply. The technologies for recovering and using steam are well developed, and in the near term this is the type of geothermal resource that will be developed most rapidly.

Geopressured Reservoirs

In geopressured reservoirs the pore fluid in a permeable formation is overpressured rather than underpressured like the natural steam reservoirs. Such reservoirs are relatively common in deep sedimentary basins. In particular, there are large areas along the Gulf Coast of Texas and Louisiana that are underlain by deeply buried sandstone beds containing highly pressurized water at moderately elevated temperatures. The water is believed to be saturated with dissolved natural gas, which, if it can be recovered economically, may contribute significantly to the currently dwindling U.S. supply.

Because the geopressured formations occur at great depths, well costs are high, and provision for recovery and pipelining of the natural gas adds to the cost of surface facilities. However, credits for the natural gas recovered may make them economic as a source. Without natural gas recovery, the heat will probably never be economically exploitable.

Research now in progress is intended to verify the temperature and gas content of known geopressured reservoirs, investigate the sustained per-well flow rates that are achievable, and determine whether or not subsidence of the land surface will follow attempts at exploitation. The resource is vast, and its natural gas content needed, so that unless the answers to these questions are disappointing, there is little doubt that the geopressured resource will be developed to some extent.

Normal-Gradient Geothermal Heat

Because the solid materials composing the upper part of the earth's crust are poor conductors of heat, a crust about 30 kilometers thick is sufficient to keep the earth's surface at an average temperature of about 15°C despite the fact that the lower crust and upper mantle may be as hot as 1000°C. However, as this insulating layer is penetrated from the surface, the temperature increases at an average rate of about 30°C per kilometer; this is commonly termed the normal geothermal gradient. Where this gradient exists, hot rock at 80°C will be found at a depth of about 2.2 kilometers, and temperatures with potential for generating electricity (above 180°C) at about 5.5 kilometers. These are accessible drilling depths, and it is estimated that, at depths less than 6 kilometers under the land mass of the United States, there exist at least 3,760,000 quads of geothermal heat at temperatures about 80°C.

In the absence of a demonstrated technology for extracting heat from this large but deep and relatively low-grade energy source, estimates of its economics and probable rate of development are speculative. However, its widespread geographical distribution suggests that it might be developed by extending heat-extraction techniques developed for use at lesser depths with hot dry rock resources (see following section), once these techniques have themselves been firmly established. For the long-range energy future of the United States, this is clearly the most important of the geothermal energy resources.

Hot Dry Rock

Particularly where the crust is thin or has recently been disturbed by volcanism or faulting, higher-than-normal geothermal gradients are often encountered. These offer the possibility of reaching a usefully high temperature with a shallower, less expensive hole. The cost advantage of doing so is sufficient so that a rather vague distinction is now commonly made between this hot-dry-rock situation and the normal-gradient one.

Extracting heat from hot dry rock requires injecting cool water through one drill hole, permitting it to circulate through either natural or induced permeable rocks until it reaches a high enough temperature, and then recovering it as either hot water or steam. Several methods of accomplishing this appear possible, and the feasibility of one of them--the use of fluid pressure to create a fracture system connecting two well bores--has recently been demonstrated in New Mexico in hot granite at a depth of about 3 kilometers. Although several important questions concerning the technology and economics of such systems have yet to be answered, it appears likely that these will be resolved within a few years and that a significant energy contribution can be expected from hot dry rock in the intermediate-range future.

Molten Lavas and Magmas

The extreme case of hot dry rock is a molten lava or magma, which may exist at a temperature higher than about 650°C, in a pool at the surface or in a reservoir contained at some depth below a recently or potentially active volcano. Aside from a few in national parks, the existence and depths of such bodies in the earth's crust are unknown, and practical means of extracting heat from them have yet to be demonstrated.

The technology needed to recover heat from a molten magma does not exist, and any energy contribution from it is extremely speculative. However, particularly because of the very high temperatures of magmas and the high efficiency with which heat at such temperatures can be used, the feasibility of extracting heat from them deserves thorough investigation.

Producibility

The uncertainty of any of the geothermal estimates, whether of a resource base or of the amount of energy potentially producible from it, is great. Inasmuch as the total resource is so extensive, any reasonable uncertainty factor could be assigned without altering the conclusion that, in the foreseeable future, the commercial production of geothermal heat will not be limited by the accessible energy supply. The constraints on its production and use will be limited by such other factors as the technology available; costs; economics; and legal, social, and environmental issues.

Although geothermal heat is in total such an enormous potential resource as to be virtually inexhaustible, individual projects are depletable. Although we have the technology to recover and are recovering energy from natural steam reservoirs, a demonstrated technology is not yet available for the other types of geothermal resources. In those cases, the full range of the drawbacks to recovering the energy is not well understood.

Much of the estimated potential geothermal energy identified by the Geothermal Resource Group (National Research Council, 1978b) consists of a thermal energy temperature range of from 80°C to 180°C and is of such low quality that it needs to be matched to the proper end uses to take full advantage of it. Geothermal heat cannot be transported very far without excessive loss of its ability to produce energy. For these reasons the use of these thermal resources should be confined to local energy needs. On the other hand, electricity is an energy form of high quality with such a full range of end uses, that capability of producing it with geothermal energy could lead to rapid development of generating capacity in the geothermal reservoir areas.

The major remaining constraints on the rapid commercial development of geothermal power are primarily institutional and relate to the difficult transition from potential to reality. Geothermal energy systems are capital-intensive, and the period between initial investment and initial return is long and, under present circumstances, uncertain. It is, of course, important that the risk of geothermal energy development as perceived by the investment community be reduced by strongly accelerated research and development and the construction and operation of pilot and demonstration plants.

In addition, the investment climate would be improved by legislation that defines and clarifies ownership of the resource, expedites leasing, coordinates and streamlines licensing and regulation, provides tax benefits for intangible costs of exploration and drilling, and provides depletion allowances similar to those already enjoyed by the petroleum and natural-gas industries.

Inasmuch as the actions necessary to accelerate geothermal energy development take time, the estimates in Table 53 show a relatively slow increase in power production in the near future followed by a rapid acceleration after about 1990 on the assumption that the economy and reliability of geothermal resources can be convincingly demonstrated.

ADVANCED ENERGY SYSTEM R&D

All the energy systems described in this section require extensive development before they can contribute to commercial energy production. Some of the technologies have not been developed at all, others require further development, and most need technical improvements to lower costs. The technologies also differ in the need for funding; some in the early stages require only moderate support for great progress, whereas others require substantial funding to support hardware development and demonstrations. These technologies offer significant prospects for meeting

Table 53 Projections for estimated installed geothermal energy
production, in quads per year

Year	Business as usual		Enhanced supply		National commitment	
	Input heat	Output electricity	Input heat	Output electricity	Input heat	Output electricity
Electricity generation						
1980	0.1	0.03	0.1	0.03	0.14	0.04
1985	0.2	0.05	0.3	0.07	0.4	0.1
1990	0.35	0.1	0.5	0.12	1.0	0.25
2000	0.8	0.2	1.4	0.4	3.1	0.8
2010	2.3	0.6	3.8	1.0	7.3	1.8
Thermal energy						
1980	0		0		0	
1985	0		0		0	
1990	0.1		0.1		0.1	
2000	0.1		0.2		0.3	
2010	0.2		0.3		1.0	

part of our future energy needs. With the exception of centralized solar electricity generation and controlled thermonuclear fusion, however, none can be expected to provide the bulk of total United States energy needs. Almost all of these technologies would require major institutional changes to contribute importantly to energy supply.

Although these systems all require research and development efforts before their potential can be realized, it should be noted that research, development, and even successful demonstration do not guarantee success. A well-ordered program should of course, weed out the concepts with low success probabilities early, before large-scale funding is needed, but even the systems carried on to advanced development have some probability of not succeeding. Indeed, even if each of a number of projects has an 80-percent probability of success, it would be necessary to carry three of them to the advanced development stage to assure a 99-percent probability that at least one will be available.

On the basis of this, the panel recommends that several advanced technologies be supported through the scientific and engineering feasibility stages, and that the most promising ones be supported into the demonstration stage.

In closing, the panel notes that the promise of a technology has little or no relation to the chances of its being successfully developed. We caution against discontinuing a development program near completion, in favor of gambling on one whose development is insufficiently advanced for problems to have been identified.

REFERENCES

National Research Council. In press. Domestic Potential of Solar and Other Renewable Energy Sources. Solar Energy Resource Group, Supply and Delivery Panel, Committee on Nuclear and Alternative Energy Systems. Supporting Paper 6. Washington, D.C.: National Academy of Sciences.

National Research Council. 1979. Alternative Energy Demand Futures. Demand and Conservation Panel, Committee on Nuclear and Alternative Energy Systems. Washington, D.C.: National Academy of Sciences.

National Research Council. 1978a. Energy Modeling for an Uncertain Future. Modeling Resource Group, Synthesis Panel, Committee on Nuclear and Alternative Energy Systems. Supporting Paper 2. Washington, D.C.: National Academy of Sciences.

National Research Council. 1978b. Controlled Nuclear Fusion: Current Research and Potential Progress. Fusion Assessment Resource Group, Supply and Delivery Panel, Committee on Nuclear and Alternative Energy Systems. Supporting Paper 3. Washington, D.C.: National Academy of Sciences.

National Research Council. 1978c. Geothermal Resources and Technology in the United States. Geothermal Resource Group, Supply and Delivery Panel, Committee on Nuclear and Alternative Energy Systems. Supporting Paper 4. Washington, D.C.: National Academy of Sciences.

7 NONENERGY RESOURCE REQUIREMENTS

This chapter treats nonenergy resource requirements: labor, capital, materials, equipment, water, and land for the design, construction, startup, operation, and maintenance of energy production facilities and associated transportation systems. These requirements are dealt with on the basis of their national and regional distributions. The framework within which critical nonenergy resource requirements were determined includes the energy production and distribution levels projected by the various resource groups of the Supply and Delivery Panel.

Quantitative estimates of the need for nonenergy resources were made using the Energy Supply Planning Model (ESPM), developed in 1974 and 1975 for the National Science Foundation and the U.S. Energy Research and Development Administration by the Bechtel Power Corporation. ESPM was used to derive the national requirements of facilities and associated resources needed to implement the selected energy development scenarios. The model simulates a well-distributed U.S. energy supply system that includes 91 types of energy extraction, processing, and transportation facilities and determines for approximately 75 categories the direct annual requirements for manpower, materials, equipment, capital, and water resources needed to design, construct, operate, and maintain those facilities. The relationships of the energy-related facilities and transport systems used in ESPM are presented in Figure 25.

Direct resource requirements in specific categories believed to be potential constraints on energy development are calculated by the model. However, indirect quantities relating support services to manpower and costs necessary to construct and operate direct energy and transportation facilities are not included in ESPM. Separate computations were performed to derive indirect nonenergy requirements and assess their impacts on alternative energy scenarios.

It is important to bear in mind that resource requirements are generated by the model without regard to the actual availability of the resources; the computer simply assumes they are all available as needed to meet scenario assumptions. Therefore, the results used here illustrate the order of magnitude of the resources needed to provide a given amount of energy but do not assess the feasibility of meeting the stated requirements.

Input for the ESPM is provided for each projection in terms of (1) the physical units for coal, petroleum, natural gas, and uranium and the number of dwellings using solar energy, (2) energy extraction requirements expressed in quads (10^{15} Btu), and (3) the electric power generating capacity in gigawatts (GWe). These input data were provided for 1975, 1985, 2000, and 2010 (Figure 26).

The lead times and project schedules for designing, building, and starting up the various facilities, and the assumptions about facility lifetime and retirement of existing facilities, are included in the computer data base. The ESPM program determines the energy-related facility additions and transportation systems necessary to supply the specified fuel mix and calculates the direct nonenergy resource requirements, which are displayed in computer printouts on a year-by-year schedule from 1977 to 2010. The computer printouts provide the following data: (1) total direct labor requirements for design, construction, startup, operation, and maintenance; (2) total direct dollar cost projections of (a) capital requirements for designing and constructing the facilities and providing materials, equipment, and utilities to operate and maintain them; and (3) natural resource requirements for the facilities in terms of fuel, land, and water consumption.

SCENARIOS

Any model is an abstraction--a simplification of reality containing little of the rich detail that characterizes energy supply, demand, and risks. The basic scenarios developed for the Supply and Delivery Panel report cover a wide range of supply and delivery patterns. Although scenarios of this type can be useful, they are inherently limited and should be viewed with great care. They are not forecasts of the future, but rather attempts to reveal the implications of a stated set of assumptions. The scenarios selected for this chapter are merely attempts to represent the limits of the practical range of the United States' energy supply and delivery capabilities over the next 35 years. They are comparable to but not identical to the scenarios used in other chapters of the Supply and Delivery Panel report.

Low Case Scenario (L)

This scenario is based on the premise that future energy policy and action will continue as in the past. No clear government policy is formulated, and no incentives beyond those already existing are provided to spur industry toward greater output of energy resources.

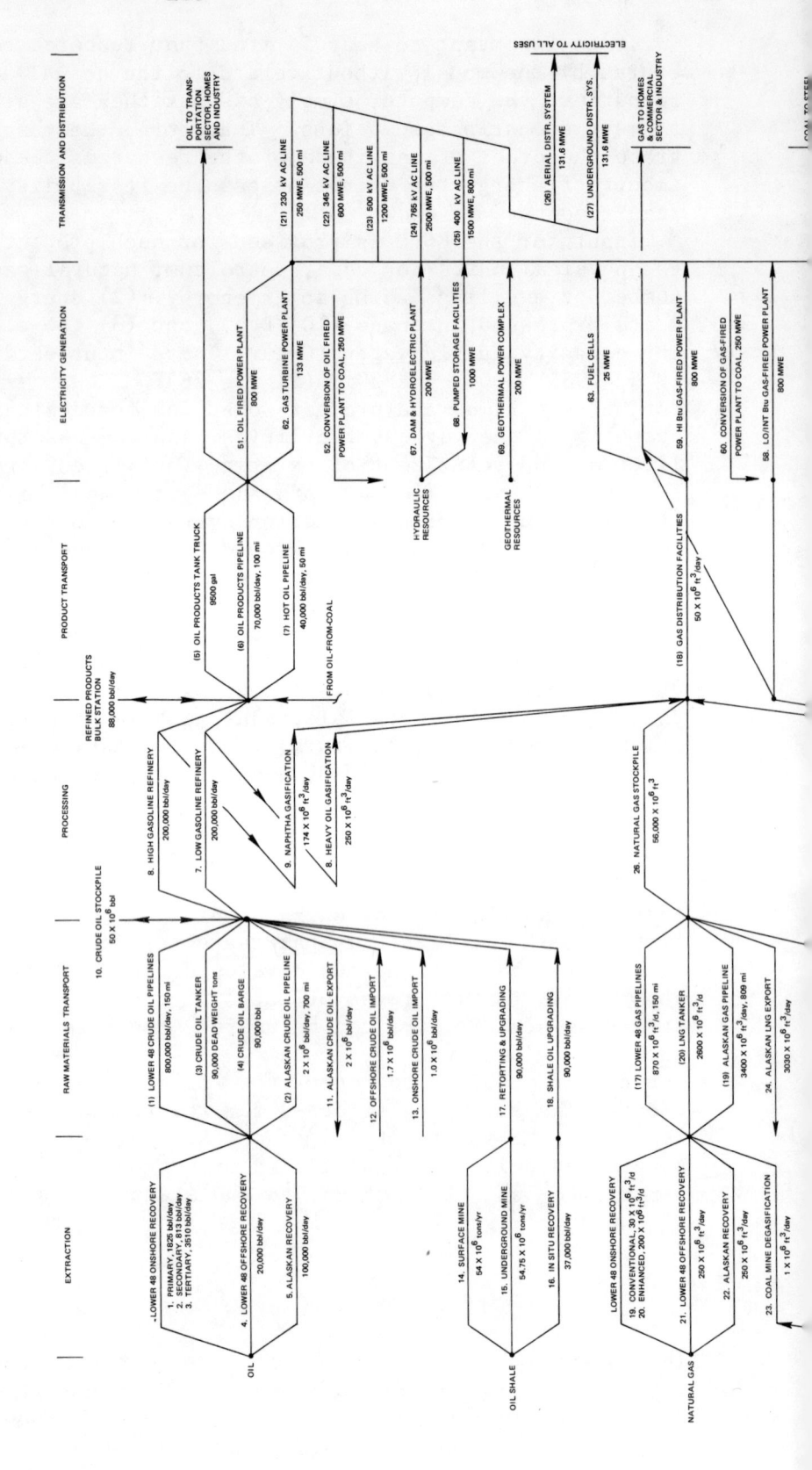

186

EXTRACTION

OIL

LOWER 48 ONSHORE RECOVERY
1. PRIMARY, 1825 bbl/day
2. SECONDARY, 813 bbl/day
3. TERTIARY, 3510 bbl/day

4. LOWER 48 OFFSHORE RECOVERY
20,000 bbl/day

5. ALASKAN RECOVERY
100,000 bbl/day

OIL SHALE

14. SURFACE MINE
54 × 10⁶ tons/yr

15. UNDERGROUND MINE
54.75 × 10⁶ tons/yr

16. IN SITU RECOVERY
37,000 bbl/day

NATURAL GAS

LOWER 48 ONSHORE RECOVERY
19. CONVENTIONAL, 30 × 10⁶ ft³/d
20. ENHANCED, 200 × 10⁶ ft³/d

21. LOWER 48 OFFSHORE RECOVERY
250 × 10⁶ ft³/day

22. ALASKAN RECOVERY
250 × 10⁶ ft³/day

23. COAL MINE DEGASIFICATION
1 × 10⁶ ft³/day

RAW MATERIALS TRANSPORT

(1) LOWER 48 CRUDE OIL PIPELINES
800,000 bbl/day, 150 mi

(3) CRUDE OIL TANKER
90,000 DEAD WEIGHT tons

(4) CRUDE OIL BARGE
90,000 bbl

(2) ALASKAN CRUDE OIL PIPELINE
2 × 10⁶ bbl/day

11. ALASKAN CRUDE OIL EXPORT
2 × 10⁶ bbl/day, 700 mi

12. OFFSHORE CRUDE OIL IMPORT
1.7 × 10⁶ bbl/day

13. ONSHORE CRUDE OIL IMPORT
1.0 × 10⁶ bbl/day

17. RETORTING & UPGRADING
90,000 bbl/day

18. SHALE OIL UPGRADING
90,000 bbl/day

(17) LOWER 48 GAS PIPELINES
870 × 10⁶ ft³/d, 150 mi

(20) LNG TANKER
2600 × 10⁶ ft³/d

(19) ALASKAN GAS PIPELINE
3400 × 10⁶ ft³/day, 809 mi

24. ALASKAN LNG EXPORT
3030 × 10⁶ ft³/day

10. CRUDE OIL STOCKPILE
50 × 10⁶ bbl

PROCESSING

8. HIGH GASOLINE REFINERY
200,000 bbl/day

7. LOW GASOLINE REFINERY
200,000 bbl/day

9. NAPHTHA GASIFICATION
174 × 10⁶ ft³/day

8. HEAVY OIL GASIFICATION
250 × 10⁶ ft³/day

26. NATURAL GAS STOCKPILE
56,000 × 10⁶ ft³

PRODUCT TRANSPORT

REFINED PRODUCTS BULK STATION
88,000 bbl/day

(5) OIL PRODUCTS TANK TRUCK
9500 gal

(6) OIL PRODUCTS PIPELINE
70,000 bbl/day, 100 mi

(7) HOT OIL PIPELINE
40,000 bbl/day, 50 mi

FROM OIL-FROM-COAL

(18) GAS DISTRIBUTION FACILITIES
50 × 10⁶ ft³/day

ELECTRICITY GENERATION

51. OIL FIRED POWER PLANT
800 MWE

62. GAS TURBINE POWER PLANT
133 MWE

52. CONVERSION OF OIL-FIRED POWER PLANT TO COAL, 250 MWE

67. DAM & HYDROELECTRIC PLANT
200 MWE
HYDRAULIC RESOURCES

68. PUMPED STORAGE FACILITIES
1000 MWE

69. GEOTHERMAL POWER COMPLEX
200 MWE
GEOTHERMAL RESOURCES

63. FUEL CELLS
25 MWE

59. HI Btu GAS-FIRED POWER PLANT
800 MWE

60. CONVERSION OF GAS-FIRED POWER PLANT TO COAL, 250 MWE

58. LO/INT Btu GAS-FIRED POWER PLANT
800 MWE

TRANSMISSION AND DISTRIBUTION

OIL TO TRANSPORTATION SECTOR, HOMES AND INDUSTRY

(21) 230 kV AC LINE
250 MWE, 500 mi

(22) 345 kV AC LINE
600 MWE, 500 mi

(23) 500 kV AC LINE
1200 MWE, 500 mi

(24) 765 kV AC LINE
2500 MWE, 500 mi

(25) 400 kV AC LINE
1500 MWE, 800 mi

(26) AERIAL DISTR. SYSTEM
131.6 MWE

(27) UNDERGROUND DISTR. SYS.
131.6 MWE

ELECTRICITY TO ALL USES

GAS TO HOMES & COMMERCIAL SECTOR & INDUSTRY

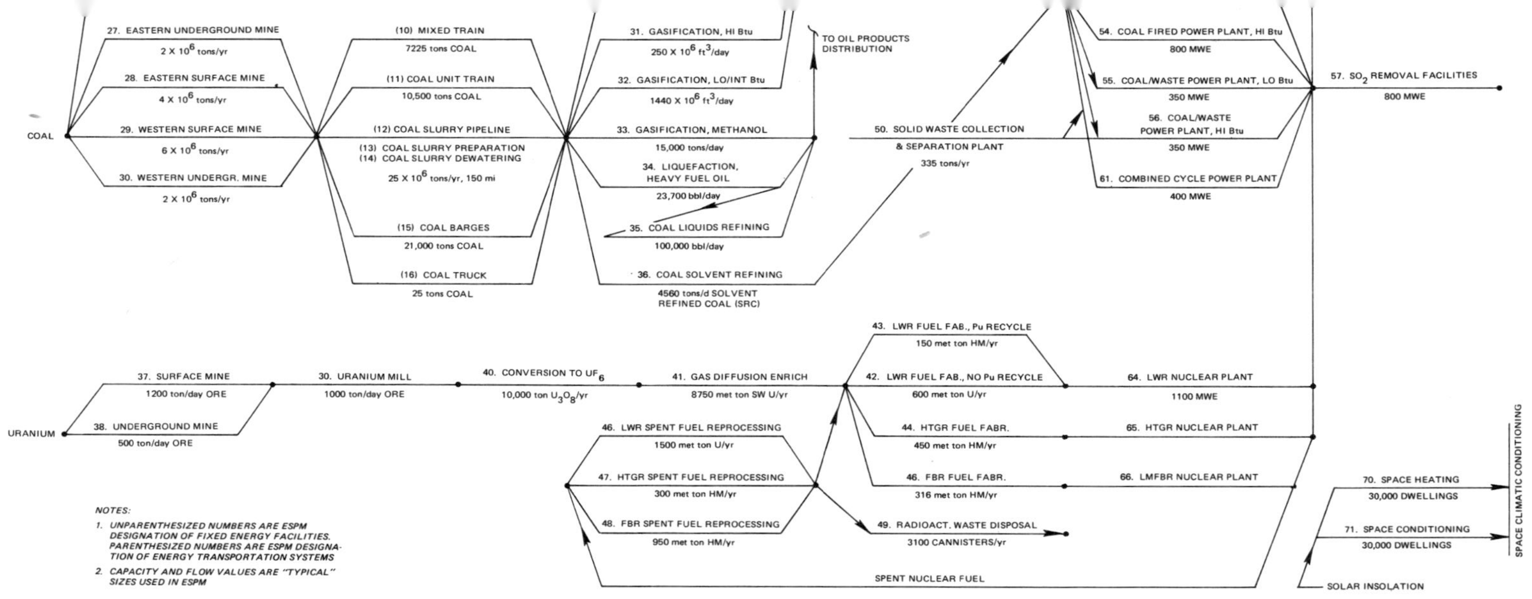

Figure 25 Flowsheet of fixed facilities and transportation systems considered in Bechtel Energy Supply Planning Model.

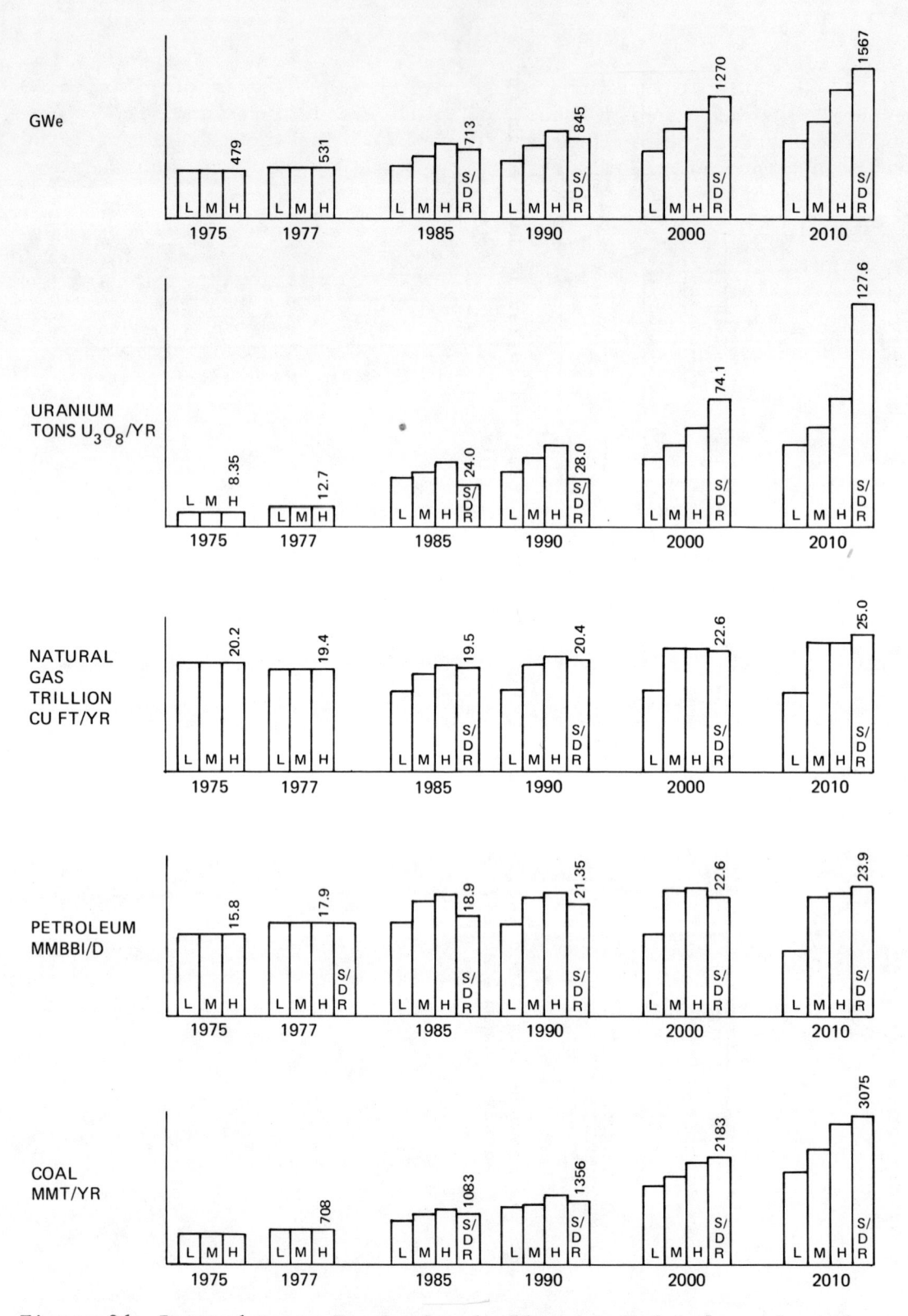

Figure 26 Input data to Energy Supply Planning Model for selected years between 1975 and 2010, in physical units.

Middle Case Scenario (M)

In this scenario it is assumed that government policy is clarified with respect to overall energy issues and plans and includes analyses on both a national and a regional basis (Figure 27). The regional computer run offers an improved analysis of interregional energy flows, transportation modes, and regional land and water limitations.

High-Intensity Electric Case Scenario (HIE)

This scenario uses the same assumptions as the Middle Case scenario except that government policy is directed toward higher generation of electricity.

Supply and Delivery Panel Recommended Case Scenario (SDR)

This scenario examines the effect of each fuel sector under favorable energy supply development. Since all energy development is not expected to be successful, this scenario attempts to identify probable upper limits.

IMPLICATIONS

The ESPM projections of energy-related facilities and their corresponding transportation systems are predicated on certain assumptions about schedules and lead times, from which the manpower, capital, land, and water needs are established. Unfortunately, the program has no way of verifying that the required facilities are being pursued within the time constraints assumed. Accordingly, if these facilities do not materialize in the real world as projected by the program, it is conceivable that the capacity will not be available when it is needed and that there will not be enough time to develop effective alternatives. It could be argued that the introduction of new technologies could avert any casualties in development. However, the introduction of a new technology takes a long time and can require enormous expenditures. Any such new technology must be available now if it is to contribute significantly to satisfying the scenario projections.

The capital requirements for the projected facilities over the next 35 years will range from $1.0 trillion to $2.8 trillion, which indicates the capital-intensive nature of the energy sector of the economy. Because of the competition for allocation of funds by other sectors of the economy, it is likely that the objectives of the Supply and Delivery scenarios may be constrained by the shortage of capital. Therefore, clear-cut government policy and adequate economic incentives will be necessary if the scenario goals are to be achieved.

The projected labor requirements appear to be attainable for timely construction of the energy-related facilities, although acceleration of the schedules might create shortages in some areas. In any event,

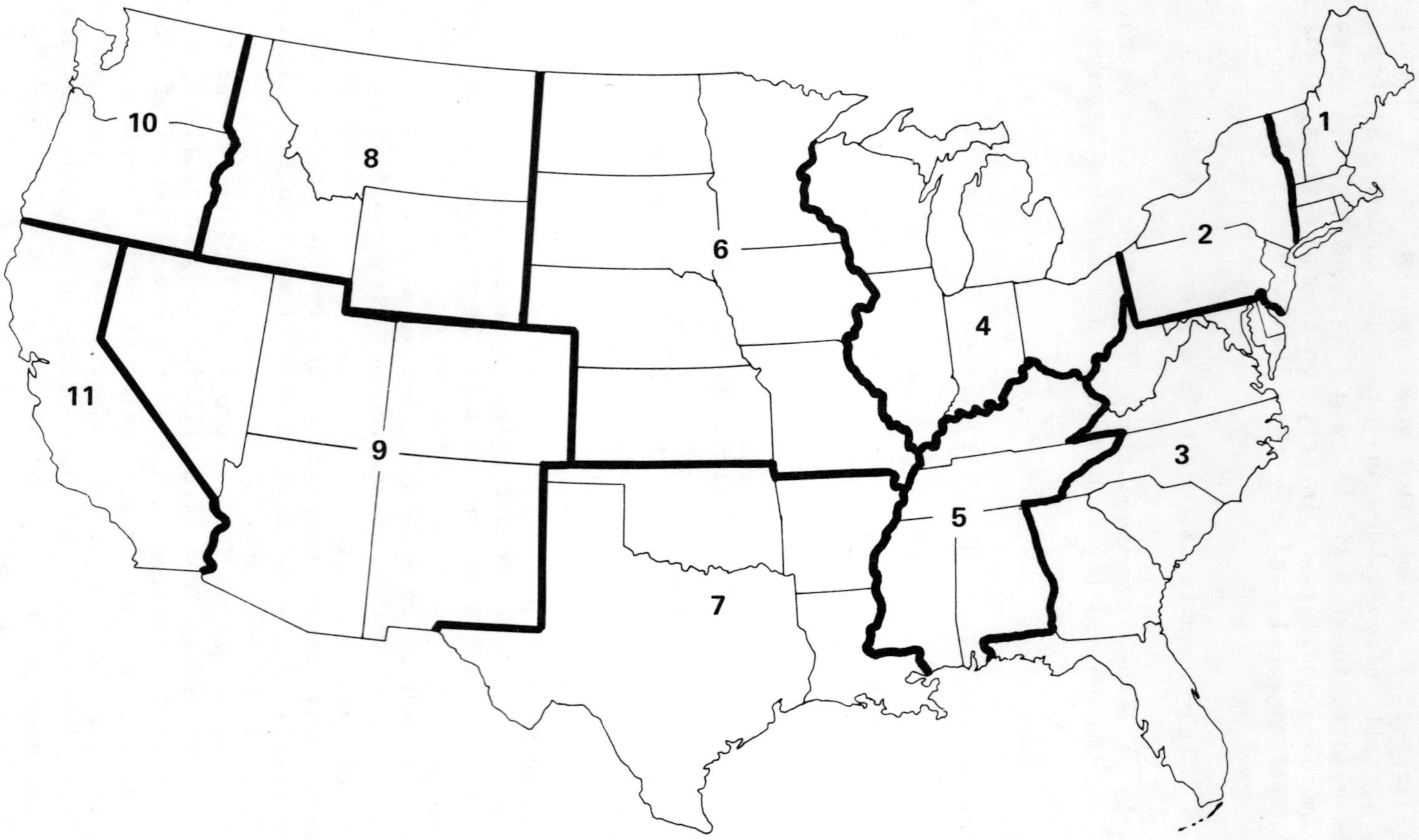

Figure 27 Regional divisions used in Energy Supply Planning Model.

with sufficient incentives, labor-saving machinery could be developed, as in coal mining now, and engineering requirements could be minimized by standardizing nuclear and fossil-fueled power plants. Such measures could help overcome any manpower deficiencies, provided that enrollment in universities and technical schools is maintained so that qualified personnel are available when needed.

The amount of land projected for the energy facilities through 2010 ranges from 164,000 to 250,000 square miles, which is about the size of the state of California. A substantial portion of the land is for underground leases and offshore oil options, however, and land does not appear likely to constrain meeting the projected goals of the scenario. Most of the problems of land use have to do with time-consuming efforts to demonstrate compliance with environmental standards. In addition, the use of land can be expanded by technology, such as applying sound principles to land restoration after surface mining.

On a national basis, it should be feasible to provide the 440 million to 670 million acre-feet of water for consumption by energy-related facilities. Unfortunately, examination of regional requirements indicates potential constraints in water-short areas of the country. Large quantities will be required for coal slurry pipelines, coal-based synthetic fuel plants, and oil shale production, all of which are likely to be located in relatively dry areas such as the Rocky Mountains and the western deserts. Incentives will be needed to promote use of dry cooling, water recycle, water pretreatment to bring it to a usable quality, and the use of two-way pipelines to move water from elsewhere and then return it to its original watershed.

Although the labor requirements and costs reported in this chapter are very large on an absolute basis, they do appear manageable in a comparative sense. The current gross national product is about $1.5 trillion, and current national employment is somewhat over 90 million. At an economic growth rate of only 2 percent per year, the gross national product in 2010 would be $3.0 trillion (1978 dollars); at 3 percent per year, it would be $4.2 trillion. The cumulative gross national product from 1977 through 2010 would be $75 trillion at a 2 percent growth rate and $92 trillion at 3 percent. Total employment by 2010 could be as high as 125 million.

On the basis of these values, capital for construction of new energy systems over the next 35 years ($1.0 trillion to $2.8 trillion) would not exceed 2.7 percent of the cumulative gross national product (2 percent growth) and could be as low as 1.1 percent (3 percent growth). During 1976, it represented about 2 percent of the gross national product; in 2010, it would account for 1.0 to 1.5 percent of the annual gross national product—not much changed from the present, assuming that the funds for energy systems construction are spent rather uniformly. Even though the comparative costs of projected new energy systems appear manageable, it must be realized that capital demands for energy systems are in competition with many alternatives uses of funds. Clearer government energy policy and provision of new economic incentives will probably be needed to reach the goals of the Middle Case, High Intensity Electric, and the Supply and Delivery Recommended Case scenarios.

Expansion of the U.S. labor pool from the 1977 figure of 90 million to 125 million by 2010 represents employment growth of 1 percent per year and gives a 34-year total of 3,623 million man-years. Growth in employment is based on two opposing trends. The low population growth rate tends to slow total employment, whereas better education of the unemployed and underemployed should increase it. The portion of the labor pool involved in design, construction, and startup of new energy facilities is currently about 0.4 percent; by 2010 it could be between 0.32 and 0.83 percent of the labor pool. Over the next 34 years, labor for providing new energy facilities will account for 0.27 to 0.78 percent of the total of 3,623 million man-years available. The number of workers needed to operate and maintain energy facilities is considerably larger and will account for 0.87 to 1.5 percent of all labor for the next 34 years. By 2010, the range will be from 0.7 of 2.1 percent of employment.

There seems little doubt that the total labor force needed to build and operate new energy facilities will be available; however, there will certainly be shortfalls in some labor categories. The most serious can be expected to be in coal mining, which has never been considered a particularly desirable occupation. To increase the number of miners from the present 150,000 to the 500,000 to 600,000 needed by 2010 will be very difficult without strong incentives.

Problems with nonenergy materials also appear slight, although there will assuredly be occasional shortages of some materials. Domestic production of pig iron was 101 million tons in the peak year of 1973 and that of raw steel 151 million tons; by the year 2000 steel output is expected to be 195 million to 260 million tons annually. Under the Supply and Delivery Panel Recommended scenario, over 323 million tons of steel will be needed for energy systems through 2010--about 10 percent of total domestic production. Because cement and aluminum have nearly infinite resource bases, there are no threats to supply of the materials, and bottlenecks could be due only to insufficient production capacity, which seems unlikely.

The supply of equipment such as large turbines, drill rigs, and heat exchange boilers will not be materials-limited, but problems may arise from time to time because of limited fabrication capacities. This could be minimized somewhat if an early procurement program is initiated for items having long lead times for delivery, since the construction of a power plant or other large energy facilities may take a decade to design and construct. The most serious problem envisioned is the possible need for crash programs in some energy sectors if the abilities to supply lag too far behind demand.

SUMMARY

The output information from the computer runs displays the direct requirements for manpower, materials, equipment, capital, land, and water related to the design, construction, startup, operation, and maintenance of the new energy facilities and their associated transport systems. The calculated values for 1976 were developed to permit comparison with 2010 requirements. Highlights of these results are summarized in Tables 28-39.

Labor

The cumulative effort between 1977 and 2010 needed to design, construct, and start up new energy facilities is projected to range from 9.9 million man-years for the Low Case scenario to 28.5 million manyears for the Supply and Delivery Panel Recommended scenario (Figure 28). The regional distribution of the cumulative manpower effort of 16.945 million man-years for the Middle Case scenario is shown in Figure 29. The annual requirements in 2010 will range from 395,000 to 1,038,000 man-years, as compared with 350,000 man-years expended in 1976. A further breakdown shows that total technical personnel (engineers, designers, supervisors, and managers) range from 73,862 to 188,243 in 2010, as compared with approximately 64,000 at present.

The cumulative effort between 1977 and 2010 necessary to operate and maintain the energy-related facilities and transport systems is projected to range from 31.5 million to 52.4 million man-years (Figure 30). Of these values the technical effort range is 4.83 million to 6.98 million man-years, the nontechnical effort is approximately 1.0 million to 2.24 million man-years, and the manual craft labor effort range is 25.6 million to 43.1 million man-years. The Middle Case regional distribution of the cumulative manpower effort of 41.225 million man-years is shown in Figure 31. The annual requirements for technical personnel in 2010 range between 163,989 and 325,166 as compared with 116,000 in 1976. Similarly, the nontechnical personnel range is between 44,085 and 132,874, as compared with 19,300 in 1976, while the manual craft labor personnel range is between 967,000 and 2,288,000, as compared with 550,000 in 1976.

Capital

The cumulative total capital needs for designing, construction, and starting up new energy facilities and their associated transportation systems between 1977 and 2010 are projected to range from $815 billion for the low scenario to $2.109 billion for the Supply and Delivery Panel Recommended Case scenario as shown in Figure 32. The corresponding additional owner costs, such as land, interest during construction, and training, range from $257 billion to $700 billion. The Middle Case regional distribution of the total cumulative capital costs of $1,382 billion is shown in Figure 33.

There will be substantial materials and equipment requirements to satisfy the needs for the construction, operation, and maintenance programs for the projected facilities. The cumulative total costs projected for material to construct and startup the facilities between 1977 and 2010 range between $133.6 billion and $402.9 billion, with equipment costs ranging between $268 billion and $504 billion.

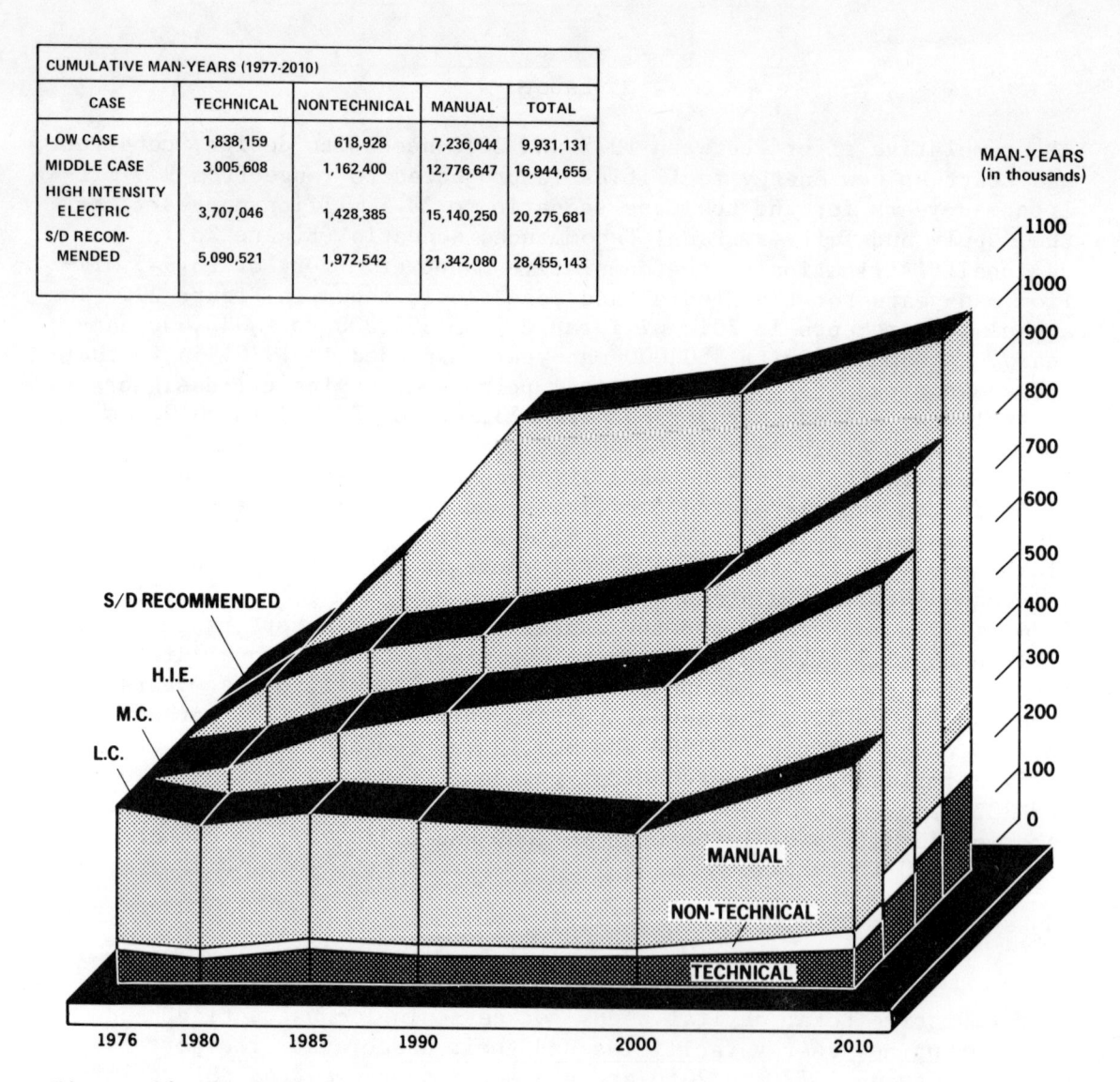

CUMULATIVE MAN-YEARS (1977-2010)				
CASE	TECHNICAL	NONTECHNICAL	MANUAL	TOTAL
LOW CASE	1,838,159	618,928	7,236,044	9,931,131
MIDDLE CASE	3,005,608	1,162,400	12,776,647	16,944,655
HIGH INTENSITY ELECTRIC	3,707,046	1,428,385	15,140,250	20,275,681
S/D RECOMMENDED	5,090,521	1,972,542	21,342,080	28,455,143

Figure 28 Projected manpower requirements for design, construction, and startup of new energy facilities from 1976 to 2010, in man-years.

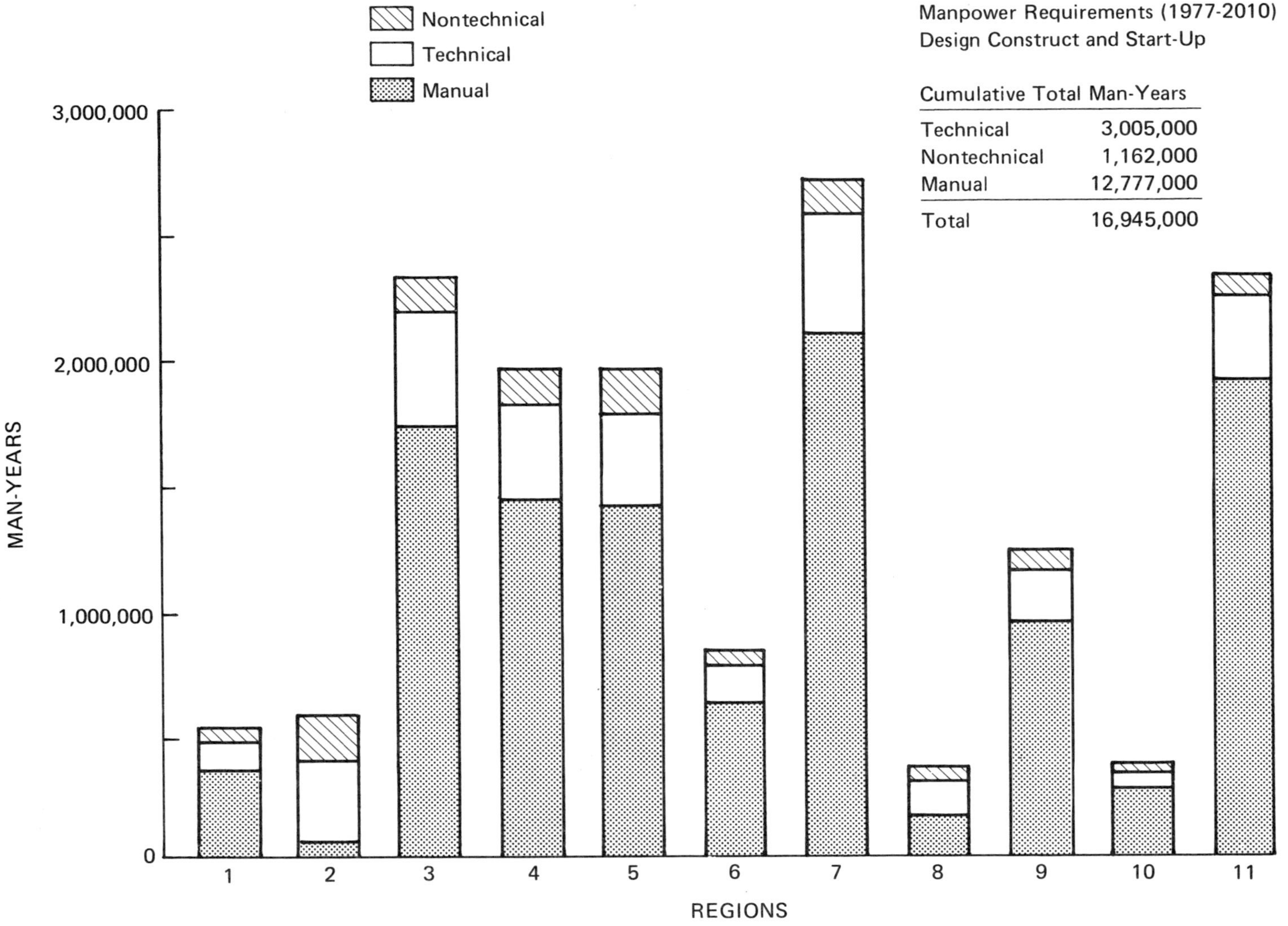

Figure 29 Projected Middle Case cumulative manpower requirements for design, construction, and startup of new energy facilities from 1977 to 2010, in man-years, by region.

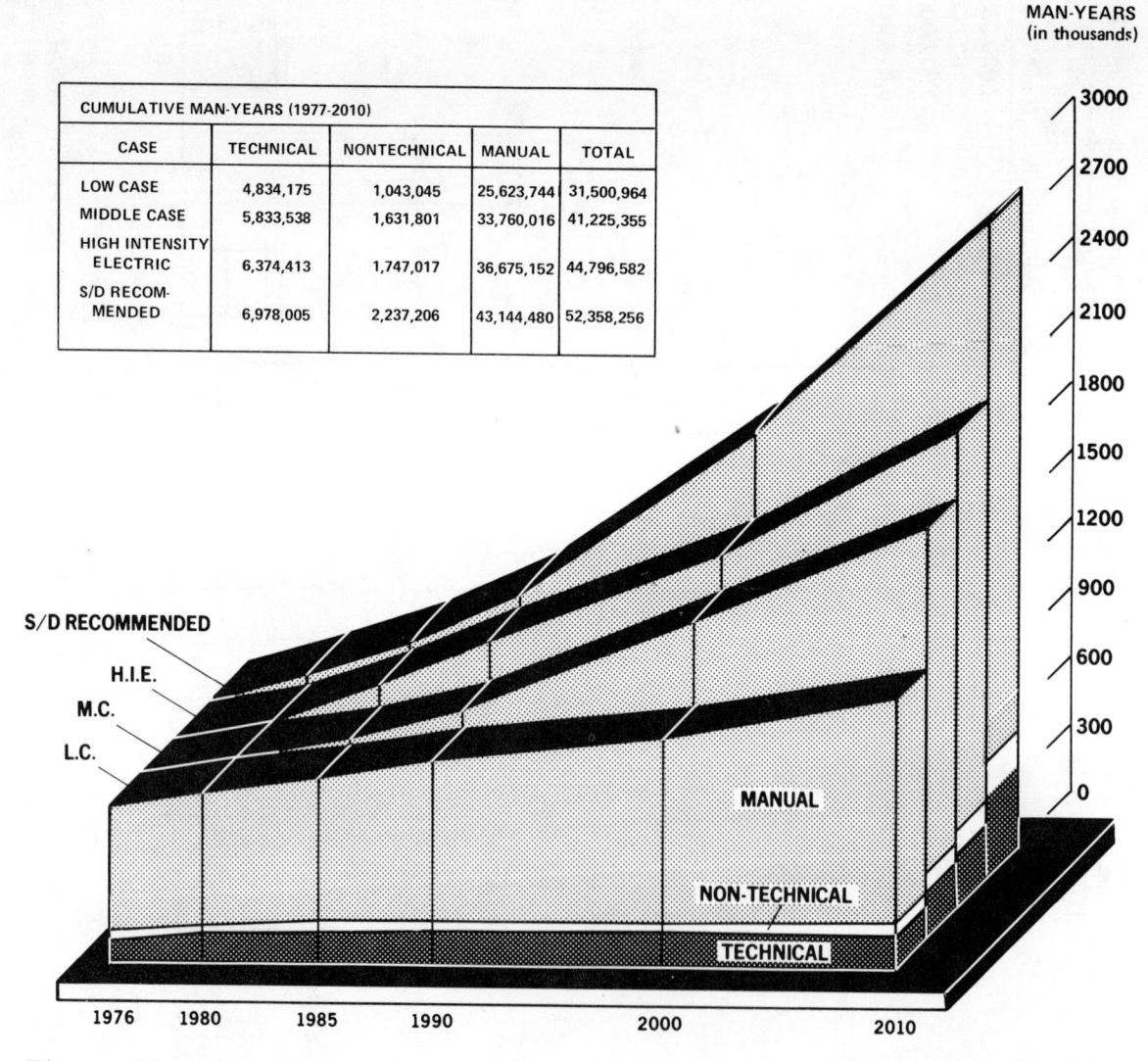

CUMULATIVE MAN-YEARS (1977-2010)				
CASE	TECHNICAL	NONTECHNICAL	MANUAL	TOTAL
LOW CASE	4,834,175	1,043,045	25,623,744	31,500,964
MIDDLE CASE	5,833,538	1,631,801	33,760,016	41,225,355
HIGH INTENSITY ELECTRIC	6,374,413	1,747,017	36,675,152	44,796,582
S/D RECOMMENDED	6,978,005	2,237,206	43,144,480	52,358,256

Figure 30 Projected manpower requirements for operation and maintenance of energy-related facilities and transport systems from 1976 to 2010, in man-years.

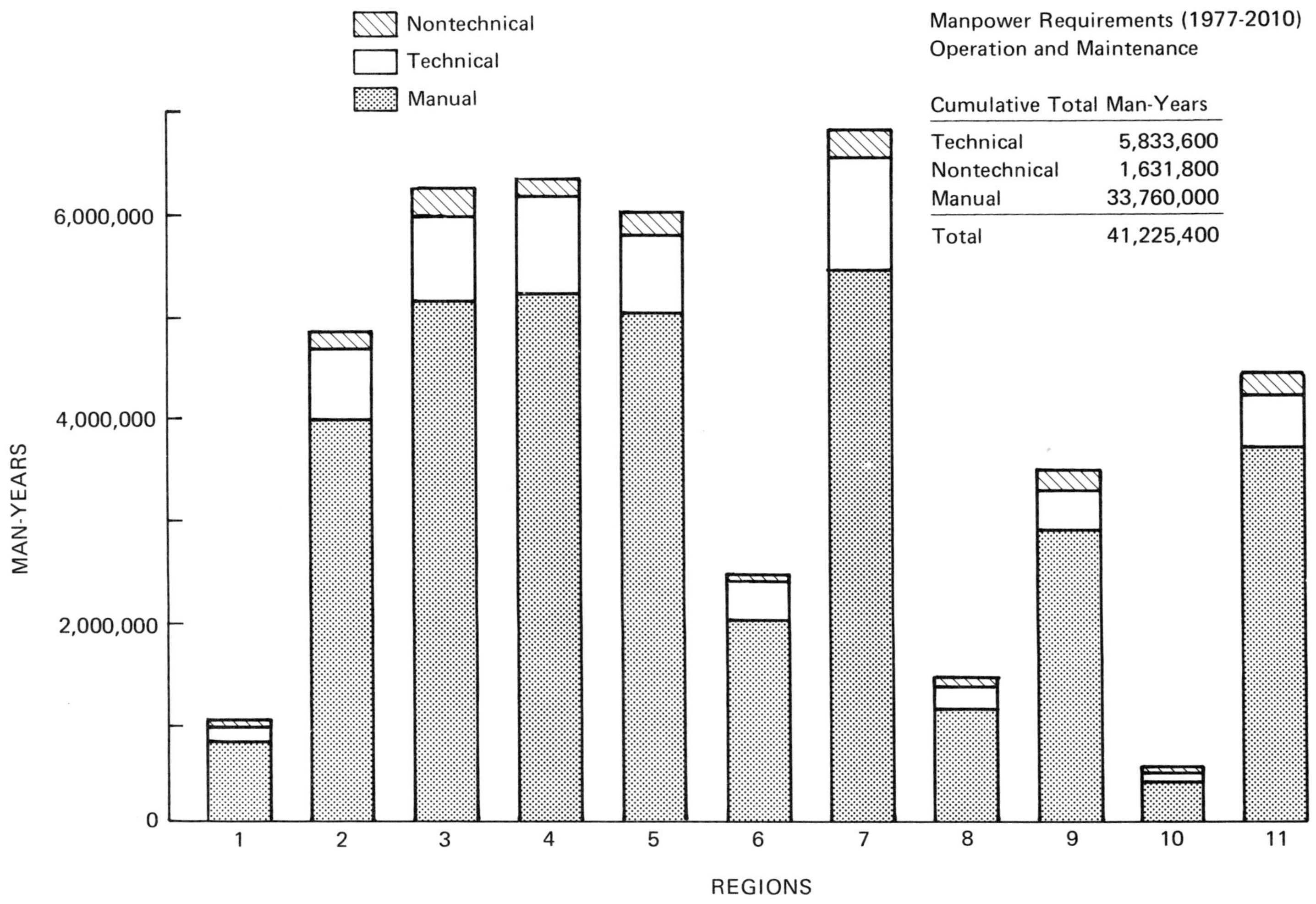

Figure 31 Projected Middle Case cumulative manpower requirements for operation and maintenance of energy-related facilities and transport systems from 1977 to 2010, in man-years, by region.

The operating and maintenance costs set forth in this section may be understated, since they do not represent the full operating costs of the energy-related facilities. In particular, the costs do not include fuel, feedstock, or manpower wages, among other factors. The cumulative total costs for materials and supplies to operate and maintain the facilities and transportation systems during the 1977-2010 period range from $259 billion to $420 billion. Similarly, total equipment costs will range from $182 billion to $331 billion, while utility costs for electricity and water will range from $112 billion to $183 billion as illustrated in Figure 34. The Middle Case regional distribution of the cumulative total costs of $790.1 billion is shown in Figure 35.

Land

The computed land requirements above ground are presented in terms of fixed land, incremental land, and right-of-way land, and those below ground in terms of underground lease land. In the period between 1977 and 2010 the projected total land needed above ground ranges from 220 million to 269 million acres, while the underground lease land is projected to range from 829 million to 1,333 million acres in Figure 36. The Middle Case regional distribution of the 248.9 million acres above ground and the 1,081 million acres below ground is shown in Figure 37.

Water

The projected cumulative and yearly average water requirements for the operation of energy facilities and their associated transportation systems are shown in Figure 38. In the period 1977-2010 the cumulative requirements will range from 439 million to 671 million acre-feet, and the yearly average from 12.9 million to 19.7 million acre-feet. The regional distribution is displayed in Figure 39.

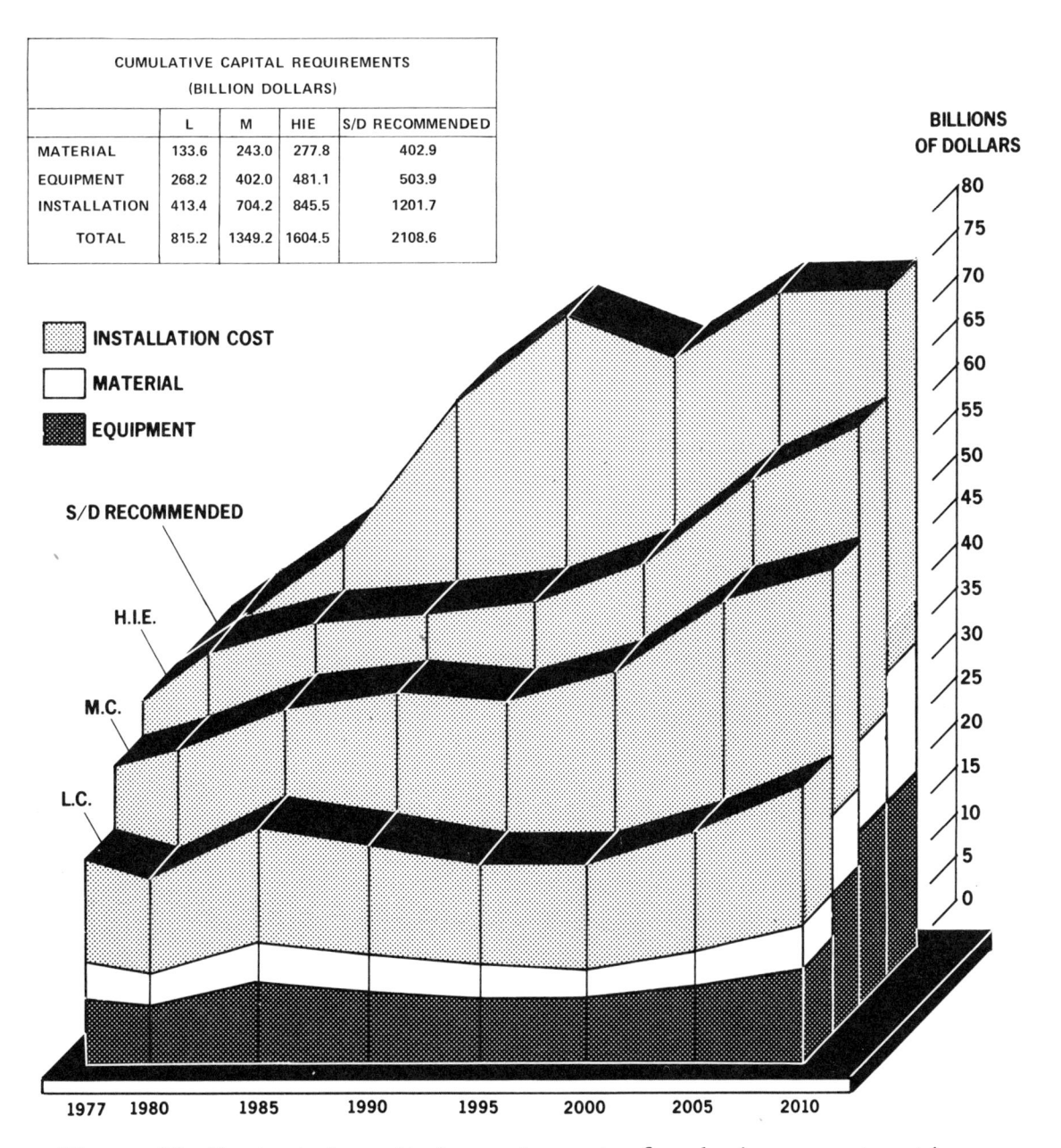

CUMULATIVE CAPITAL REQUIREMENTS (BILLION DOLLARS)				
	L	M	HIE	S/D RECOMMENDED
MATERIAL	133.6	243.0	277.8	402.9
EQUIPMENT	268.2	402.0	481.1	503.9
INSTALLATION	413.4	704.2	845.5	1201.7
TOTAL	815.2	1349.2	1604.5	2108.6

Figure 32 Projected capital requirements for design, construction, and startup of new energy facilities from 1977 to 2010, in billions of dollars.

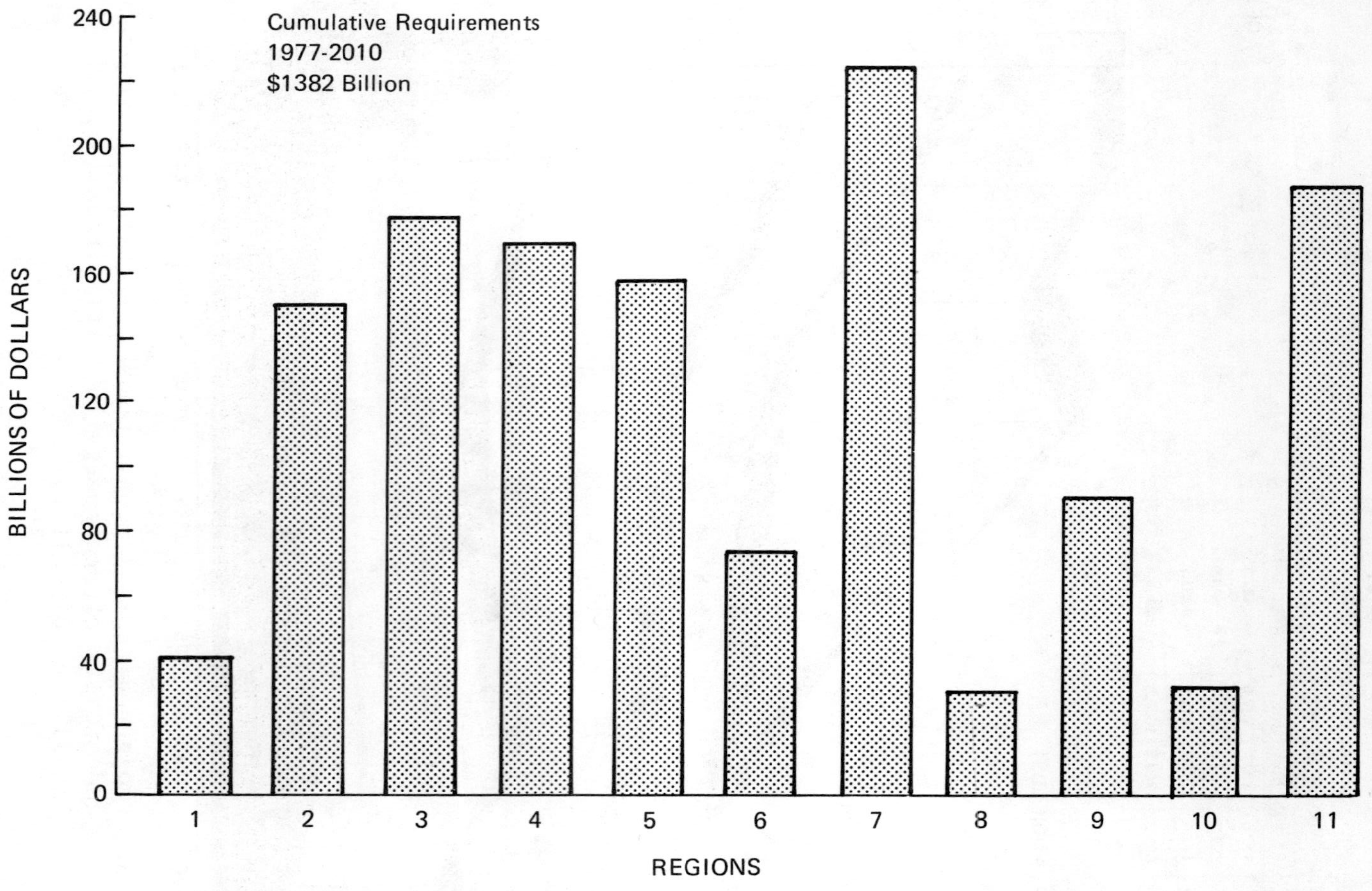

Figure 33 Projected Middle Case cumulative capital requirements for design, construction, and startup of new energy facilities from 1977 to 2010, in billions of dollars, by region.

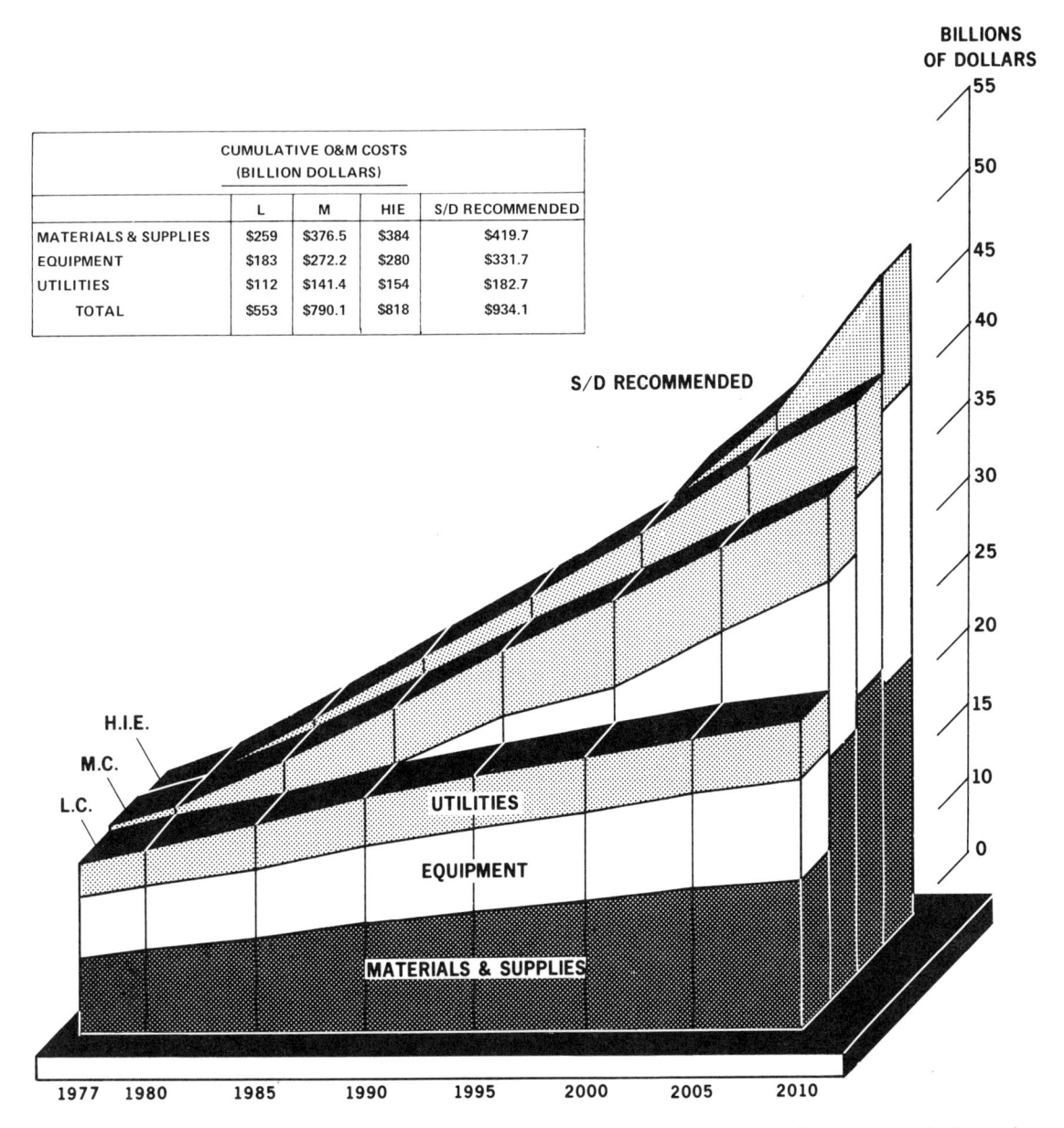

CUMULATIVE O&M COSTS (BILLION DOLLARS)				
	L	M	HIE	S/D RECOMMENDED
MATERIALS & SUPPLIES	$259	$376.5	$384	$419.7
EQUIPMENT	$183	$272.2	$280	$331.7
UTILITIES	$112	$141.4	$154	$182.7
TOTAL	$553	$790.1	$818	$934.1

Figure 34 Projected operation and maintenance costs for material and equipment resources to operate and maintain energy-related facilities and transport systems from 1977 to 2010, in billions of dollars.

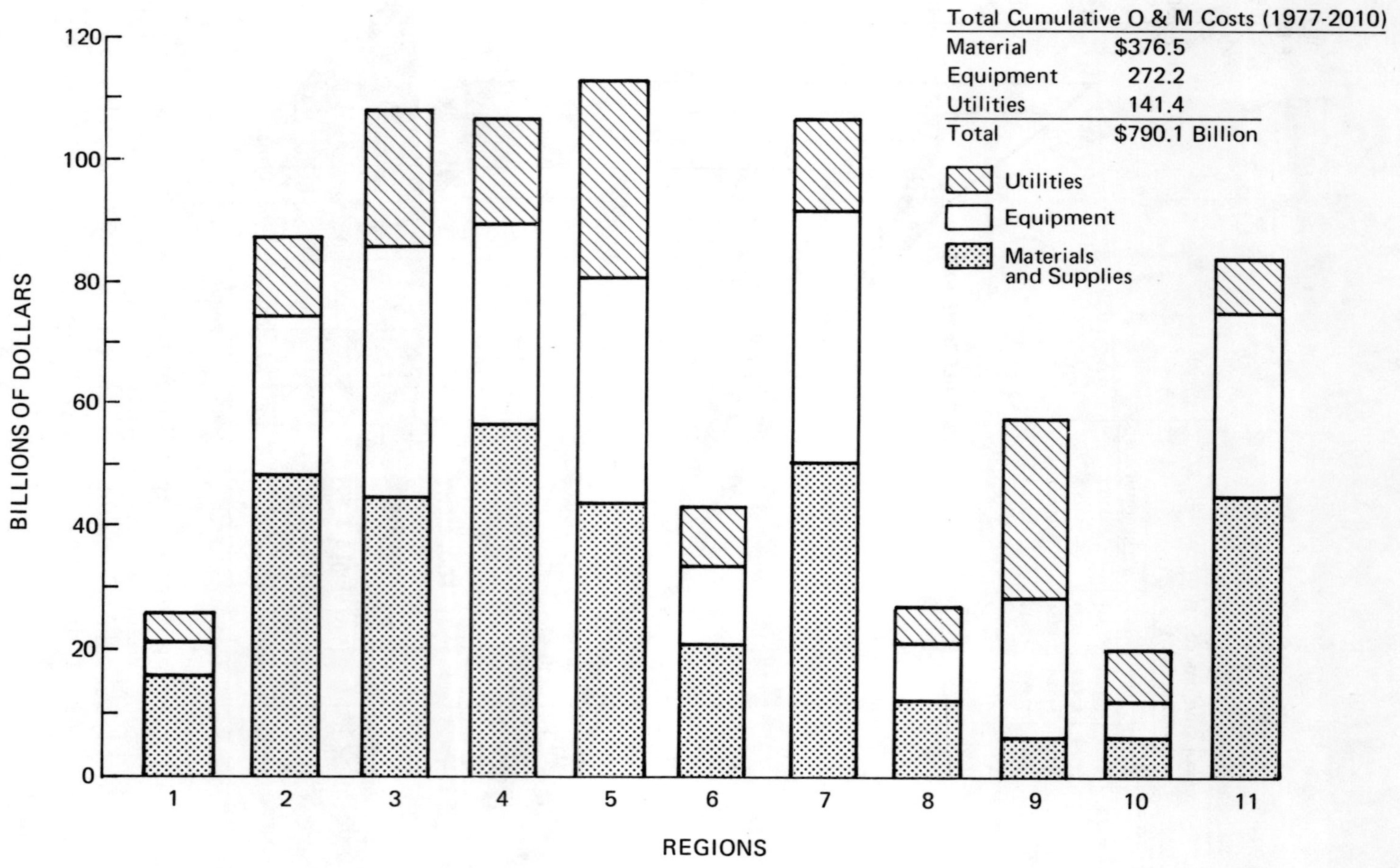

Figure 35 Projected Middle Case cumulative operation and maintenance costs for energy-related facilities and transport systems from 1977 to 2010, in billions of dollars, by region.

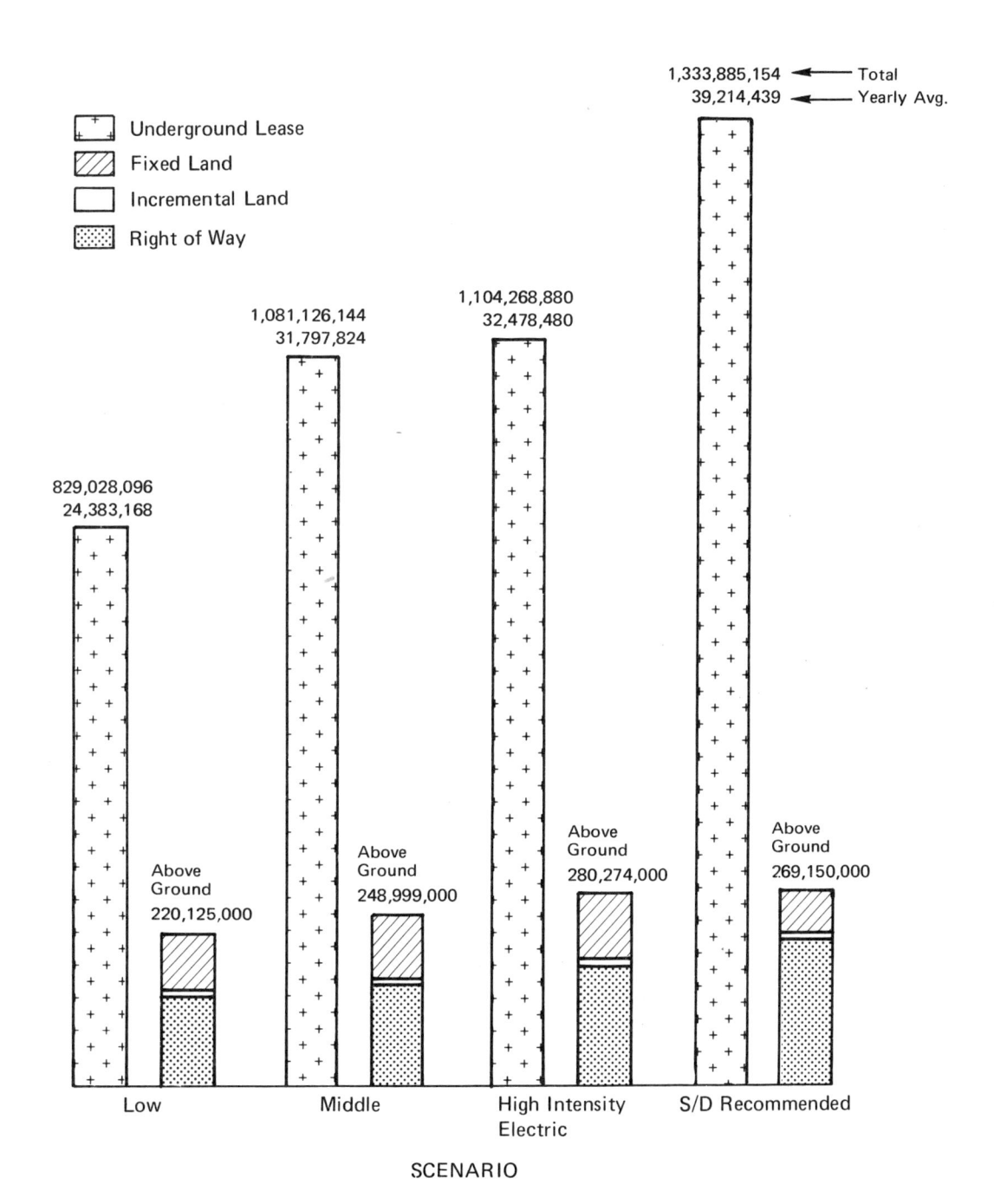

Figure 36 Projected cumulative land requirements from 1977 to 2010, in acres.

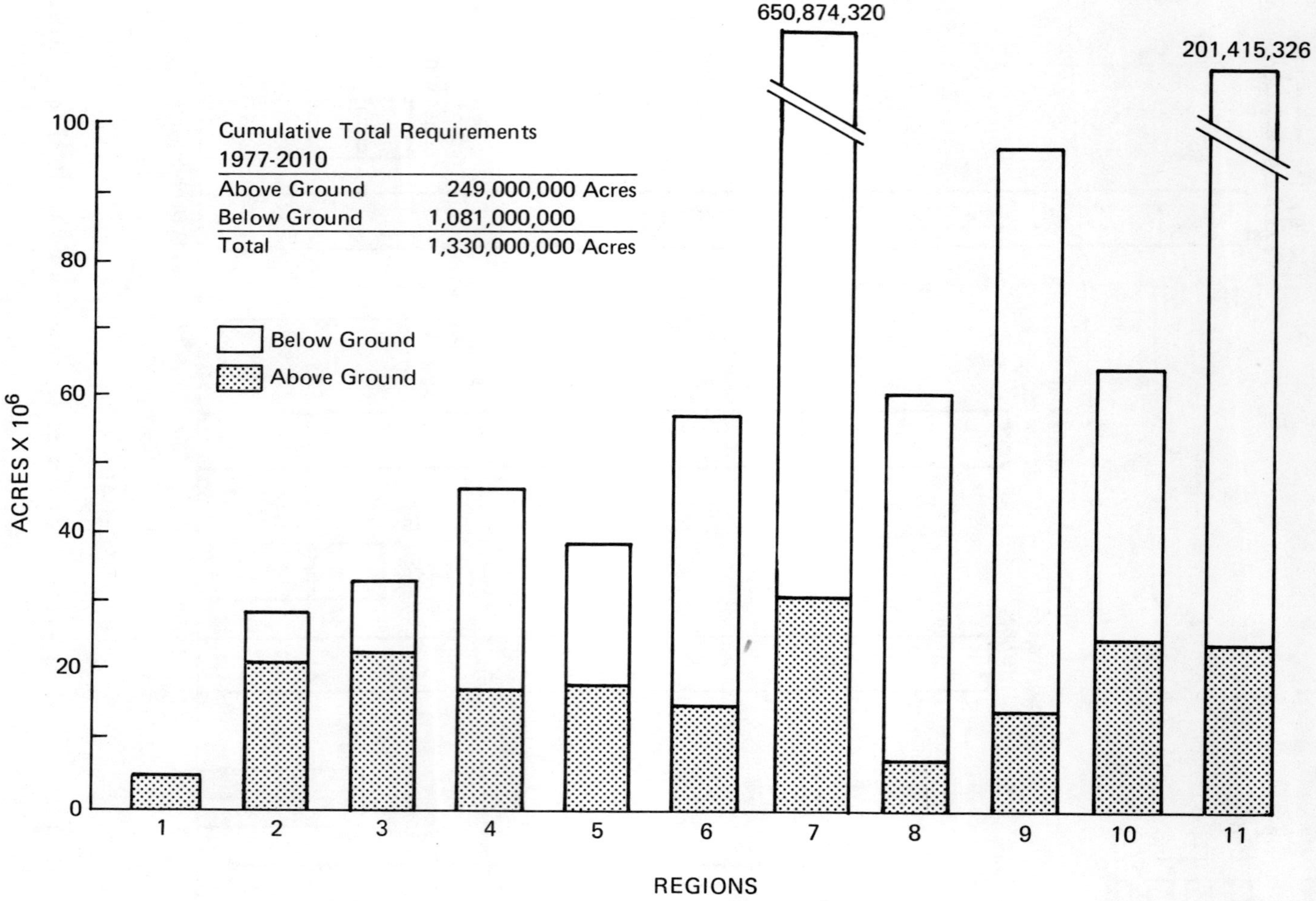

Figure 37 Projected Middle Case cumulative land requirements from 1977 to 2010, in acres, by region.

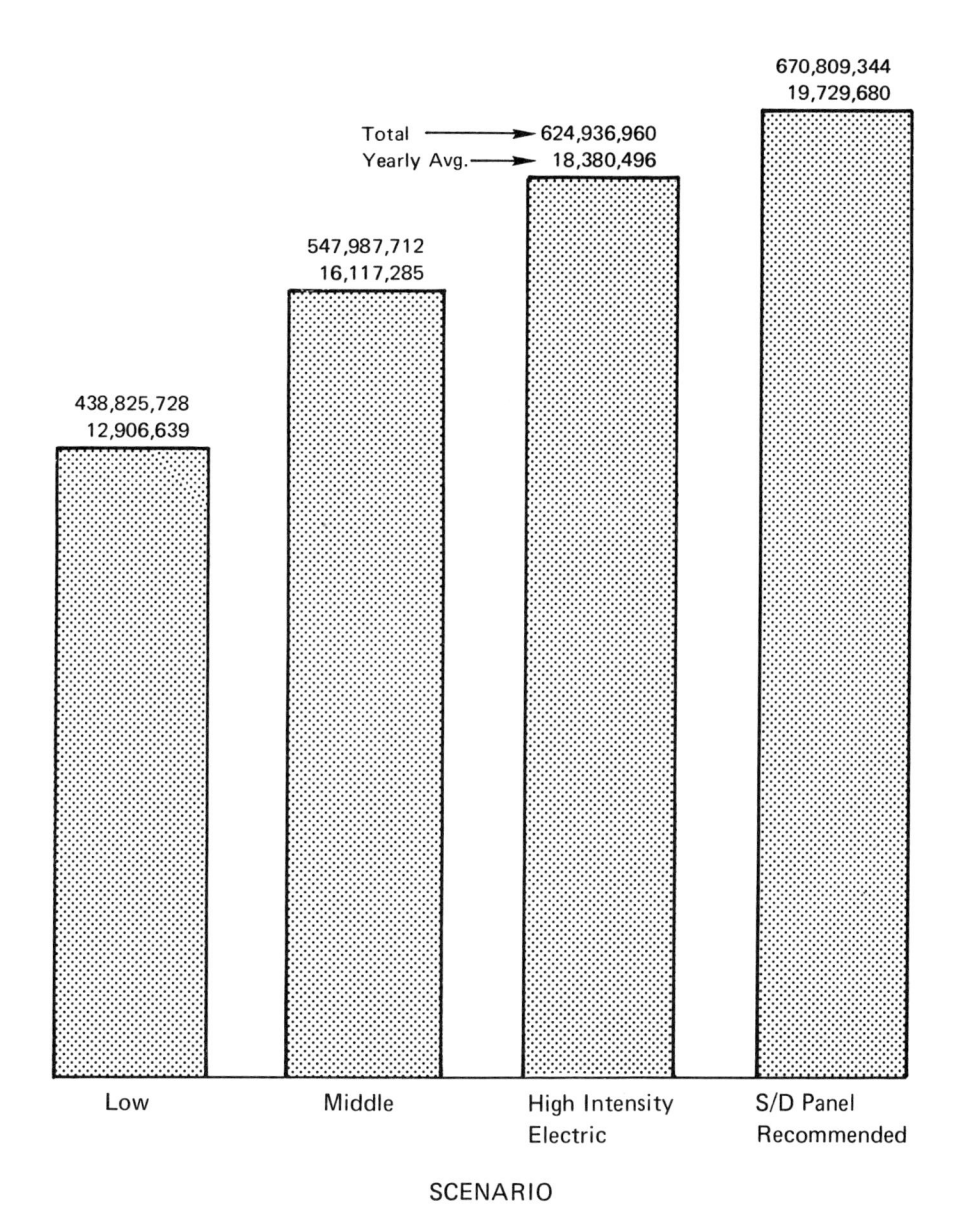

Figure 38 Projected cumulative water consumption requirements from 1977 to 2010, in acre-feet.

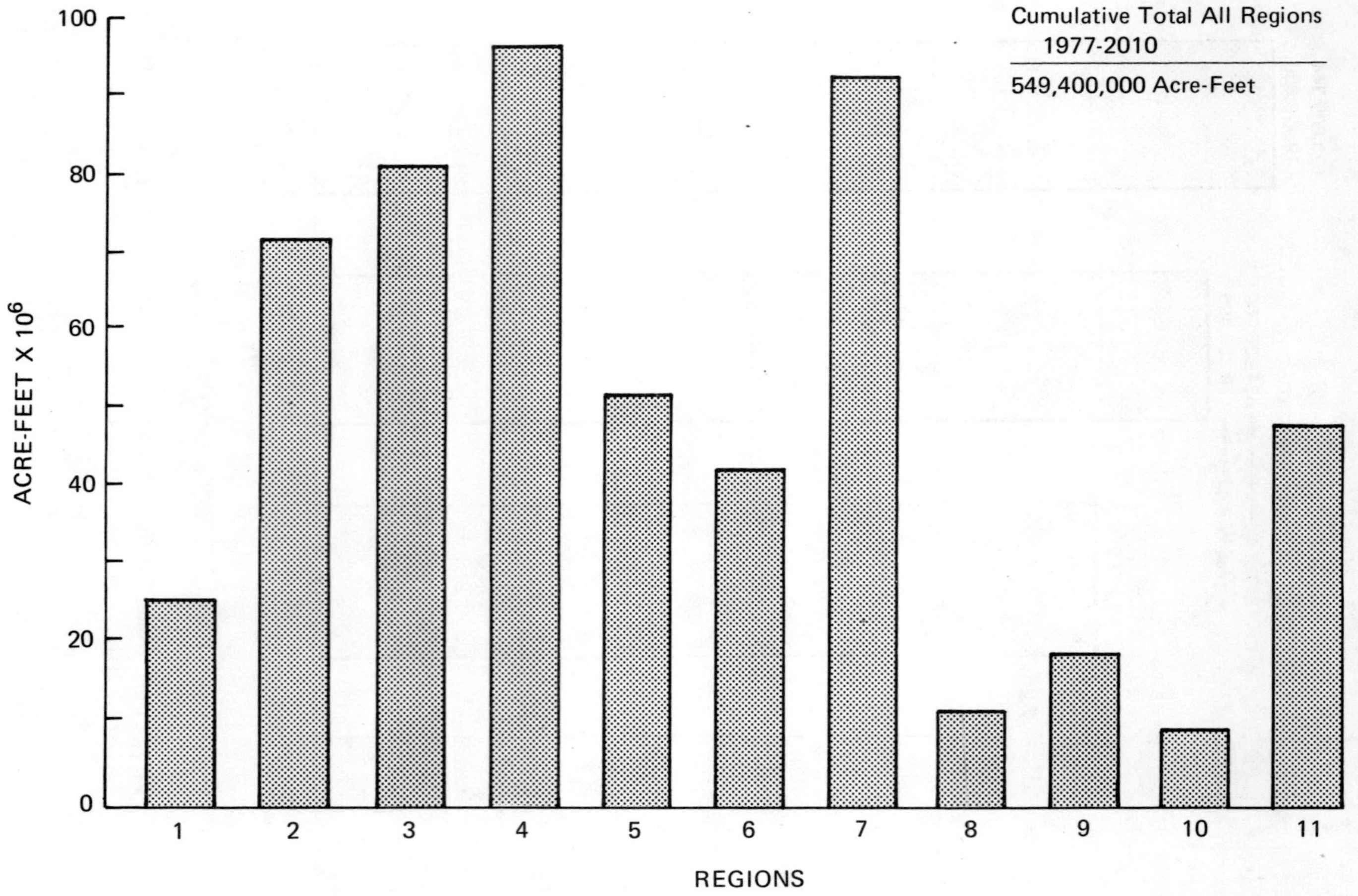

Figure 39 Projected Middle Case cumulative water consumption requirements from 1977 to 2010, in acre-feet, by region.

APPENDIX: GLOSSARY OF TECHNICAL TERMS

Accelerator (particle accelerator): A device for imparting large kinetic energy to electrically charged elementary particles such as electrons, protons, deuterons, and helium ions through the application of electrical and/or magnetic forces. Common types of particle accelerators are direct voltage accelerators, cyclotrons, betatrons, and linear accelerators.

Actinides: A group name for the series of radioactive elements from element 89 (actinium) through element 103 (lawrencium). The series includes uranium and all the man-made transuranic elements.

Bioconversion: The conversion of organic wastes into methane (natural gas) through the action of microorganisms.

Binary Cycle: An energy recovery system based on the transfer of heat from one fluid (e.g., hot brine from a geothermal well) to a second fluid (e.g., pure water or an organic liquid) from which the heat is ultimately extracted and used.

Blanket: A layer of fertile material such as uranium-238 or thorium-232 that is placed around the reactor core. Its major function is to produce fissile isotopes from fertile blanket material.

Boiling-Water Reactor (BWR): A light water reactor that employs a direct cycle; the water coolant that passes through the reactor is converted to high-pressure steam that flows directly through the turbines.

Breeder reactor: A nuclear reactor that produces more fissile material than it consumes. In fast breeder reactors high-energy (fast) neutrons produce most of the fissions, while in thermal breeder reactors fissions are principally caused by low-energy (thermal) neutrons.

Breeding ratio: The ratio of the number of fissionable atoms produced in a breeder reactor to the number of fissionable atoms consumed in the reactor. The _breeding gain_ is the breeding ratio minus one.

Btu (British thermal unit): The amount of energy necessary to raise the temperature of one pound of water by one degree Fahrenheit, from 39.2 to 40.2 degrees Fahrenheit.

Capacity factor: The ratio of the amount of product (e.g., electrical energy or geothermal brine) actually produced by a given unit, system, or plant per unit of time to its maximum production rate. Also called "load factor."

Cogeneration: The generation of electricity with direct use of the residual heat for industrial process heat or for space heating.

Combined cycle: A combination of a steam turbine and a gas turbine in an electrical generating plant, with the gas turbine exhaust heat used in raising steam for the steam turbine.

Conversion ratio: The ratio of the number of atoms of new fissionable material produced in a converter reactor to the number of atoms of fissionable fuel consumed. See "breeding ratio."

Converter reactor: A reactor that produces some fissionable material, but less than it consumes. In some usages, a reactor that produces a fissionable material different from the fuel burned, regardless of the ratio. In both usages the process is known as conversion.

Curie: A measure of intensity of the radioactivity of a substance; i.e., the number of unstable nuclei that are undergoing transformation in the process of radioactive decay. One curie equals the disintegration of 3.7×10^{10} nuclei per second, which is approximately the rate of decay of one gram of radium.

Depletion allowance: A tax credit based on the permanent reduction in value of a depletable resource that results from removing or using some part of it.

Dry hot rock (geothermal): See "hot dry rock."

Fertile material: A material, not itself fissionable by thermal neutrons, which can be converted into a fissile material by irradiation in a reactor. There are two basic fertile materials, uranium-238 and thorium-232. When these materials capture neutrons, they are partially converted into plutonium-239 and uranium-233, respectively.

Flashing: The rapid change in state from a liquid to a vapor without visible boiling, resulting usually from a sudden reduction in the pressure maintained on a hot liquid.

Fluidized bed: A body of finely divided particles kept separated and partially supported by gases blown through or evolved within the mass, so that the mixture flows much like a liquid.

Fly ash: Fine solid particles of noncombustible ash entrained in the flue gases arising from the combustion of carbonaceous fuels. The particles of ash may be accompanied by combustible unburned fuel particles.

Fuel cell: A device that produces electrical energy directly from the controlled electrochemical oxidation of fuel. It does not contain an intermediate heat cycle, as do most other electrical generation techniques.

Fuel cycle: The various processing, manufacturing, and transportation steps involved in producing fuel for a nuclear reactor, and processing fuel discharged from the reactor. The uranium fuel cycle includes uranium mining and milling, conversion to UF_6, isotopic enrichment, fuel fabrication, reprocessing, recycle of recovered fissile isotopes, and disposal of radioactive wastes.

Gas centrifuge process: A method of isotopic separation in which heavy gaseous atoms or molecules are separated from light atoms or molecules by centrifugal force.

Gaseous diffusion: A process used to enrich uranium in the isotope uranium-235. Uranium in the form of a gas (UF_6) is forced through a thin porous barrier. Since the lighter gas molecules containing uranium-235 move at a higher velocity than the heavy molecules containing uranium-238, the lighter molecules pass through the barrier more frequently than do the heavy ones, producing a slight enrichment in the lighter isotope. Many stages in series are required to produce material enriched sufficiently for use in a light water reactor.

Geopressured reservoir (geothermal): A hydrothermal reservoir in which the pore fluid is confined under pressure significantly greater than normal hydrostatic pressure, developed principally by the weight of overlying rocks and sediments. Also called "overpressured" and "geopressurized" reservoirs.

Geothermal gradient: The rate at which the temperature of the earth increases with depth below its surface. This varies widely from place to place, but the average or "normal" geothermal gradient is typically about 30°C per kilometer of depth (16.5°F/1000 ft.).

Heavy water: Water containing significantly more than the natural proportion (one in 6500) of heavy hydrogen (deuterium) atoms to ordinary hydrogen atoms. Heavy water is used as a moderator in certain reactors because it slows down neutrons effectively and also has a low cross-section for absorption of neutrons.

High-level waste: A byproduct of the operation of nuclear reactors that includes a variety of aqueous wastes from fuel reprocessing and their solidified derivatives, such as aqueous waste, alkaline aqueous waste, calcine, crystallized salts, insoluble precipitates, salts of cesium and strontium extracts, and coating wastes from chemical decladding of fuel elements.

High-temperature gas-cooled reactor (HTGR): A graphite-moderated, helium-cooled advanced converter reactor that utilizes the thorium fuel cycle. The initial core is fueled with a mixture of fully enriched U-235 and thorium. When operated in the recycle mode, the reactor is refueled with a mixture of U-233 (produced from thorium) with the balance of the fissile material provided from an external source of fully-enriched U-235.

Hot dry rock (geothermal): Naturally heated but unmelted rock sufficiently low in either permeability or pore-fluid content so that wells drilled into it do not yield either hot water or steam at commercially useful rates. To be compared with hydrothermal reservoirs.

Hydrothermal reservoir: A body of porous, permeable rock, gravel, or soil containing natural steam or naturally heated water at a temperature significantly above the average temperature at the earth's surface.

Isotope: One of two or more atoms with the same atomic number (i.e., the same chemical element) but with different atomic weights. Isotopes usually have very nearly the same chemical properties, but somewhat different physical properties.

Kerogen: A solid, largely insoluble organic material, occurring in oil shale, which yields oil when it is heated in the absence of oxygen.

Light water reactor (LWR): A nuclear reactor that uses ordinary water as both a moderator and a coolant, and utilizes slightly enriched uranium-235 fuel. There are two commercial light water reactor types--the boiling water reactor (BWR) and the pressurized water reactor (PWR).

Liquefied natural gas (LNG): Natural gas cooled to $-259°F$ so that it forms a liquid at approximately atmospheric pressure. As natural gas becomes liquid it reduces volume nearly 600-fold, thus allowing economical storage and making long-distance transportation economically feasible. Natural gas in its liquid state must be regasified and introduced to the consumer at the same pressure as other natural gas. The cooling process does not alter the gas chemically and the regasified LNG is indistinguishable from other natural gases of the same composition.

Liquefied petroleum gas (LPG): A gas containing certain specific hydrocarbons which are gaseous under normal atmospheric conditions, but can be liquefied under moderate pressure at normal temperatures. Propane and butane are the principal examples.

Load factor: Capacity factor (q.v.).

Low-level waste: Generally a solid by-product of special nuclear materials production, utilization, and research and development. Examples of solid low-level waste are discarded equipment and materials, filters from gaseous waste cleanup, ion exchange resins from liquid waste cleanup, liquid wastes that have been converted to solid form by techniques such as mixing with cement, and miscellaneous trash. Low-level liquid waste is generally decontaminated and released under controlled conditions.

Milling (uranium processing): A process in the uranium fuel cycle in which ore that contains only about 0.2 percent uranium oxide (U_3O_8) is concentrated into a compound called yellowcake, which contains 80 to 90 percent U_3O_8.

Moderator: A material such as ordinary water, heavy water, or graphite, which is used in a reactor to slow down high-velocity neutrons, thus increasing the likelihood of further fission.

Nuclear waste: The radioactive products formed by fission and other nuclear processes in a reactor. Most nuclear waste is initially in the form of spent fuel. If this material is reprocessed, new categories of waste result: high-level, transuranic, and low-level wastes (and others).

Particulates: Microscopic pieces of solids that emanate from a range of sources and are the most widespread of all substances that are usually considered air pollutants. Those between 1 and 10 microns are most numerous in the atmosphere, stemming from mechanical processes and including industrial dusts, ash, etc.

Plutonium: A heavy, radioactive, man-made metallic element with atomic number 94, created by absorption of neutrons in U-238. Its most important isotope is Pu-239, which is fissionable.

Pressurized water reactor (PWR): A light-water moderated and cooled reactor that employs an indirect cycle; the cooling water that passes through the reactor is kept under high pressure to keep it from boiling, but it heats water in a secondary loop that produces steam that drives the turbine.

Primary containment: An enclosure which surrounds a nuclear reactor and associated equipment for the purpose of minimizing the release of radioactive material in the event of a serious malfunction in the operation of the reactor.

Pyrolsis: Decomposition of materials through the application of heat with insufficient oxygen for complete oxidation.

Quad: A quantity of energy equal to 10^{15} British thermal units.

Reactor core: The central portion of a nuclear reactor containing the fuel elements and the control rods.

Reprocessing: A generic term for the chemical and mechanical processes applied to fuel elements discharged from a nuclear reactor; the purpose is to recover fissile materials such as plutonium-239, uranium-235, and uranium-233 and to isolate the fission products

Reserves: Resources that are known in location, quantity, and quality and that are economically recoverable using currently available technologies.

Resource (energy): That part of the resource base which is believed to be recoverable using only current or near-current technology, without regard to the cost of actually recovering it. To be distinguished from both "resource base" and "reserve" (q.v.).

Resource base (energy): The total quantity of energy or of any given energy-producing or energy-related material that is estimated to exist in or on the earth or in its atmosphere, independent of quality, location, or the engineering or economic feasibility of recovering it.

Scrubber: An air pollution control device that uses a liquid spray for removing pollutants such as sulfur dioxide or particulate matter from a gas stream by absorption or chemical reaction.

Secondary recovery: Methods of obtaining oil and gas by the augmentation of reservoir energy, often by the injection of air, gas, or water into a production formation (see tertiary recovery).

Solar constant: The solar radiation falling on a unit area at the outer limits of the earth's atmosphere.

Spectral shift reactor: A reactor in which a mixture of light water and heavy water is used as the moderator and coolant. The ratio of light to heavy water is varied to change (shift) the energy spectrum of the neutrons in the reactor core. Since the probability of neutron capture varies with neutron velocity, a measure of reactor control is thus obtained.

Synthesis gas: A fuel gas containing primarily carbon monoxide and hydrogen; it can be used after careful removal of impurities, particularly sulfur compounds, for conversion to methane (high Btu gas), methanol, liquid hydrocarbons, and a wide variety of other organic compounds.

Tailings: Waste material from a separation process. Commonly the finely divided waste from a mineral-separation operation.

Tails: Contraction of "tailings" (q.v.).

Tails (or tailings) assay: The percentage of valuable material that remains unrecovered in the tailings of a separation process.

Tar sands: Hydrocarbon-bearing deposits distinguished from more conventional oil and gas reservoirs by the high viscosity of the hydrocarbon, which is not recoverable in its natural state through a well by ordinary production methods.

Tertiary recovery: Use of heat and methods other than air, gas, or water injection to augment oil recovery (presumably occurring after secondary recovery).

Thorium: A radioactive element of atomic number 90; naturally occurring thorium has one main isotope--thorium 232. The absorption of a neutron by a thorium atom can result in the creation of the fissile material uranium-233.

Throwaway fuel cycle: A fuel cycle in which the spent fuel discharged from the reactor is not reprocessed to recover residual plutonium and uranium values.

Transuranic elements: Radioactive nuclides generated as fission products from the fissioning of nuclear fuel during reactor operation and as induced activity from the capture of neutrons in fuel cladding, reactor structures, and reactor coolant.

Uranium: A radioactive element of atomic number 92. Naturally occurring uranium is a mixture of 99.28% U-238, 0.71% U-235, and 0.0058% U-234. U-235 is a fissile material and is the primary fuel of light water reactors. When bombarded with slow or fast neutrons, it will undergo fission. U-238 is a fertile material which is transmuted to Pu-239 upon the absorption of a neutron.

Uranium hexafluoride (UF_6): A compound of uranium, which is used in gaseous form in the enrichment of uranium isotopes.

Yellowcake: A uranium concentrate which results from the milling (concentrating) of uranium ore. It typically contains 80-90 percent uranium oxide.